P9-AGS-299

The Nature of Atoms

ALAN HOLDEN

The Nature of Atoms

1971
Oxford University Press
New York and Oxford

©1971 by Oxford University Press
Library of Congress Catalogue Card Number: 71-167752
Printed in the United States of America

General Preface

THIS monograph is one of a connected series of three by the same author: *Stationary states*, *The nature of atoms*, and *Bonds between atoms*. Like its fellows, it was written for the Conference on New Instructional Materials in Physics, sponsored by the Commission on College Physics with the support of the National Science Foundation, and held at the University of Washington in the summer of 1965.

The purpose of the conference was to create effective ways of presenting physics to college students who are not preparing to become professional physicists. Such an audience might include prospective secondary-school physics teachers, prospective practitioners of other sciences, and those who wish to learn physics as one component of a liberal education.

The form of these, and of other monographs originating at that conference, implements a principle of experimental pedagogy promulgated there. Each monograph is 'multi-level': each consists of several sections arranged in order of increasing sophistication. Their authors hope that such papers can be fragmented for use in existing formal courses, or can be associated with other papers to provide the basis for new kinds of courses.

Contents

FOREWORD

1 THE ATOMIC HYPOTHESIS 1

Outlines the history of the idea that matter is made of atoms, mentions some criticism that has been levelled at that hypothesis, and suggests its status in present scientific thinking.

2 THE CHEMICAL ELEMENTS 7

Describes the development and character of the position held by the elementary species of atoms in today's picture of matter, introducing especially the chemical reasoning leading to the periodic table.

3 MODELS OF AN ATOM 12

Summarizes the growth of the pictures of an atom as an assembly of particles, describing the ideas of Thomson, and of Rutherford and Bohr, and introducing today's wave-mechanical ideas.

4 A PARTICLE IN A ONE-DIMENSIONAL WELL 25

Examines the wave-mechanical behaviour of a particle confined to a one-dimensional square well that has a top, providing by this artificial example some useful insights into the more complicated behaviour of an electron in an atom.

5 A PARTICLE IN A CENTRAL FIELD 35

Derives the permissible angular dependences of the wave functions for a particle moving in a field of force with spherical symmetry, of which the Coulomb force exerted by an atomic nucleus on an electron is the most important example.

6 ANGULAR MOMENTA OF A PARTICLE IN A CENTRAL FIELD 44

Shows that each sort of angular dependence derived in the preceding chapter belongs to a precise value of the total orbital angular momentum of the electron, and examines the conditions under which a precise value can be given also to one component of that momentum.

7 HYDROGENLIKE ATOMIC STATES 51

Derives the radial dependence of the wave functions for the states of an electron bound to an atomic nucleus, and the associated levels of energy, completing the description of the exact results available for discussing atomic behaviour.

8 THE BUILDING-UP PRINCIPLE 60

Makes qualitative use of the results of the preceding three chapters to show how the approximation by one-electron wave functions to the description of a many-electron atom can explain the atomic behaviour summarized by the periodic table.

9 THE TWO-ELECTRON DELTA-WELL ATOM 68

Illustrates the difficulties of pursuing exact wave-mechanical methods in discussing a system of many interacting particles, by applying those methods to a simple system of two particles in one dimension and comparing the calculated energy with the measured energy of a helium atom.

10 MANY-ELECTRON ATOMS 78

Deals with the wave functions of the electrons in atoms more quantitatively than Chapter 8, and less exactly than the attempt in Chapter 9, by imagining each electron to move in a central field of force exerted both by the nucleus and by a spherical cloud of the other electrons.

TABLES 88

INDEX 91

Foreword

A GREAT diversity of observations compels the inference that matter is made of atoms, despite the fact that nobody has ever 'seen' one. Over the past two centuries the atomic hypothesis has been especially fruitful in providing explanations for the behaviour of matter. This monograph summarizes the history of that hypothesis up to the beginning of this century, when atoms themselves were found to be divisible into electrons and nuclei.

From that point, the monograph confines its attention to the behaviour of the electrons in atoms, ignoring the fact that the nuclei have in turn been found to be divisible. There is much justification for separating the study of nuclear structure from that of electronic behaviour. The phenomena by which the two sorts of structure can be investigated experimentally are quite different, and are characterized by absorptions and evolutions of energy that fall in widely different energy ranges. Almost all the phenomena of ordinary life are controlled by the electronic behaviour of the participating atoms, and differences of nuclear structure have little effect on that behaviour.

The wave-mechanical discussion in the last seven chapters of this monograph presupposes some acquaintance with wave mechanics. The relevant ideas are summarized in another monograph, *Stationary states*. In turn this monograph provides information that will give help in reading the monograph *Bonds between atoms*.

ALAN HOLDEN

The Nature of Atoms

1. The Atomic Hypothesis

FOR most people the difficulty of living is not abated by adding to it a concern about how the world is made. But there have always been some for whom that concern has been unavoidable—for a few, even joyful—and they have constantly remade the world of the others. Today a conspicuous result of their efforts is the notion that all matter is made of atoms; it is a notion held by everybody who has heard of it, whether or not he can cite any evidence or authority for the idea.

The story of how this notion arose is one of the most interesting and difficult studies in intellectual history. Shreds of the idea appear in ancient Hindu philosophy, particularly in the writings of Kanada. More definitely it emerges as one of the many competing doctrines of Greek philosophy in the fifth century B.C.; and its principal Greek proponents, Leucippus and Democritus, are often credited with originating it. Four centuries later Lucretius expounded it in some detail in his poem *De Rerum Natura.*

But what was the substance of those early thinkers' vision of the world? Its interpretation is obscured now by problems of the authenticity of documents and the meaning of words. The very existence of Leucippus has been questioned, and his ideas and those of Democritus are known primarily through the writings of their philosophical opponents. Democritus, combating the proposition that 'anything which can be done once can be done again', seems to have taught that the repeated division of matter, at any rate, cannot be carried on indefinitely. He visualized the ultimately indivisible particles as infinite in kind, but made of a single dense material. The differences in kind resided mostly in their shapes and their sequences of juxtaposition with others. Thus the particles of water, smooth and round, are able to roll over one another, whereas the jagged particles of iron hook on to one another to form a solid body.

Can we know today how strongly Democritus held to such concrete pictures, how much he was led to overspecify them by the techniques of Grecian controversy, how often his opponents are only reporting a straw man knocked down? And can we estimate well the relative amounts of observation on the one hand, and of free speculation on the other, that formed the views of the early Greek atomists? The ancient scene provided most of the same gross features as the modern to suggest an atomic constitution of matter. Then, as now, solid salt would disappear in water and reappear when the water evaporated. Surely someone would suggest that perhaps the salt is not destroyed and then recreated but instead is dispersed into invisible fragments and then reassembled.

Some have regretted that the doctrines of the atomists lost the contest in the rough-and-tumble of Greek philosophical controversy. In any case, soon after the time of the atomists, philosophical attention began to look more inward into man and less outward into the external world. The early robust enthusiasm for explaining nature gave way to introspective examination of how man can know, cultivating the view, always latent in Greek thinking, that man can discover the universe by thought alone. As a result of this evolution, the legacy of Greek philosophy became dominated by two towering figures, Plato and Aristotle.

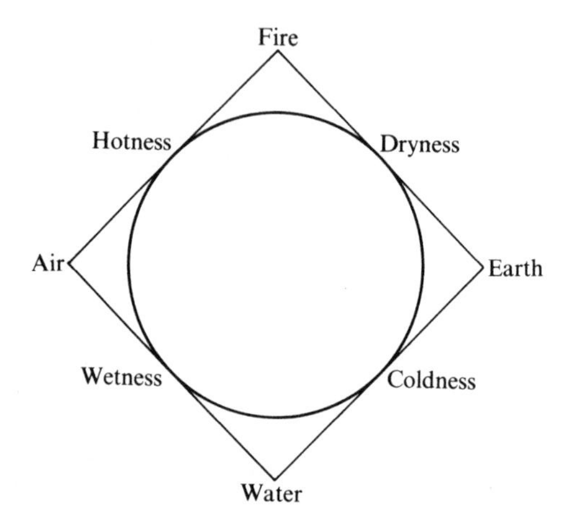

FIG. 1.1. Aristotle taught that the different forms of matter resulted from the impress, on a neutral matter-stuff, of four formative principles: hotness and dryness, and their opposites, coldness and wetness. In particular, pairs of these principles produced the 'four elements': fire, earth, water, and air.

The latter in particular, by giving some of his widely diversified attention to the question of how matter is made, bequeathed the picture suggested in Fig. 1.1, which directed further thinking in both the Christian and the Arab worlds for many centuries. The Arabs, especially, cultivated what they believed must be the practical consequences of Aristotle's doctrine. If all the kinds of matter result from the impress of formative principles on a single substance, one kind should be transformable into another. At the base of Arabian alchemy lay the search for ways of imposing in practice the formative principles suited to particular transformations.

At least two separate questions are now distinguishable in the early controversies: (1) whether matter is made of separate particles or a continuous smear, and (2) whether it is made of a single underlying stuff or many stuffs. Descriptions were constructed by combining answers to these questions in many ways; and the task of deciphering sympathetically the meanings of those descriptions is often impeded by misconceptions coming from the slowly changing connotations of ancient words over centuries of use.

'Element', for example, is a word used differently in describing the modern view of matter and Aristotle's view, for clearly in Aristotle's world there was only one element in the present sense of the word: the neutral stuff of which everything, including 'the four elements', is composed. And during the long life of the Aristotelian doctrine, the primacy of those four especially primitive 'elements' declined. With the development of metallurgy by the Arabs and the growth of pharmacology at the hands of the iatrochemists, new emphases led to new 'elements'. By the sixteenth century the *tria prima* of Paracelsus, 'mercury', 'sulphur', and 'salt', held sway.

The overturn of this way of looking at the world was accomplished in the following century when Robert Boyle proposed in *The Sceptical Chymist* that, while there are indeed primitive substances, there are more than three. 'I mean by elements ... certain Primitive and Simple or perfectly unmingled bodies; which not being made of other bodies, or of one another, are the Ingredients of which all those call'd perfectly mixed Bodies are immediately compounded, and into which they are ultimately resolved.' All other substances are formed from these, he proposed, not by formative influences but by the conjunction of their particles in various combinations, in which the particles of the primitive substances remain and from which they can be recaptured.

There followed a lively search for the primitive substances and the rules by which they combine. Clearly implicit in that search was the rebirth of an atomic theory in a pluralistic form—a theory of indestructible particles of many primitive substances—and the succeeding 150 years saw rapid progress in developing the new ideas. The study of gases in particular supported the notion that matter comes in small particles. Isaac Newton suggested in 1686 that a gas will expand when it can because it is made of tiny particles that repel one another. In 1738, Daniel Bernoulli laid the foundation for the present kinetic theory of gases by showing that the connection between the volume and the pressure of a gaseous sample, discovered by Boyle and by Edme Mariotte independently, can be explained by a model (Fig. 1.2) in which small particles of the gaseous substance are continually rushing about, bombarding the walls of their container.

By the beginning of the nineteenth century, investigation of the properties of matter and the chemical transformations it sustains had accumulated a body of facts ripe for synthetic interpretation. It was John Dalton, son of a poor weaver, whose insight first fitted the pieces together. In *A New System of Chemical Philosophy* (1808–10), he wrote,

'The existence of ultimate particles of matter can scarcely be doubted, though they are probably much too small even to be exhibited by microscopic improvement. I have chosen the word "atom" to signify those ultimate particles, in preference to particle, molecule, or any other diminutive term,

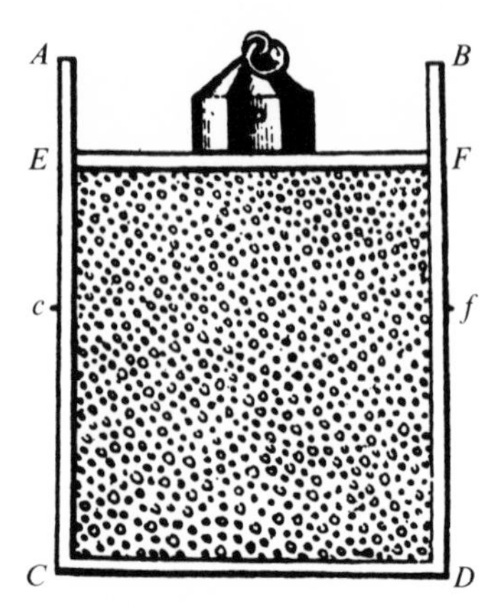

FIG. 1.2. Using this mental model of a gas, Daniel Bernoulli showed in 1738 that the relation between the pressure of a gas and its volume could be explained by supposing that its constituent particles are in constant motion, beating against the walls of the vessel that confines them.

because I conceive it is much more expressive; it includes in itself the notion of indivisible,† which the other terms do not ... . Chemical analysis and synthesis go no further than to the separation of particles one from another, and to their reunion. No new creation or destruction of matter is within the reach of chemical agency.'

Apparently Dalton was first attracted toward these conclusions by contemplating the implications of meteorological observations that he had made over many years, and by further study of the absorption of various gases by water. Ascribing definite relative weights to the different species of atoms, he tested his idea that different substances result from combining such atoms in simple integral ratios by applying it to the reported chemical analyses of various substances, and found agreement. Dalton's own quantitative work was rather rough and ready; some of it, indeed, was far from right. The most celebrated chemist of his day, Sir Humphry Davy, though an admirer of Dalton, said that he was a 'very coarse experimenter' who 'almost always found the results he required, trusting to his head rather than his hands'. Dalton's insight was surely brilliant; his picture of substances (Fig. 1.3) as made of tiny molecules, each composed of a few elementary atoms that remain essentially the same in any of their occurrences, still directs the interpretation of chemical results today.

In spite of its many successes, a reaction against a too-great concreteness of the atomic hypothesis arose among some philosophers and physicists near the end of Dalton's own century. Positivists argued against accepting an object that cannot be seen, erecting a barrier to all natural knowledge

† Dalton here invokes the Greek etymology of the word: *a* (not) plus *temnein* (to cut), whence *atomos* (the indivisible).

that purports to transcend the limits of resolution and wavelength sensitivity characteristic of the human eye. Recalling the aesthetic principle of the sciences that limits hypotheses to no more than are necessary to explain the facts, Hans Bucherer likened the atomic hypothesis to a scaffold, which can be taken down when the building is finished. Hardly more hospitably, Gustav Hertz called it a picture or model that we can make of phenomena, since they behave in many respects as if matter consists of atoms.

To be sure, the work of our own century has shown that an atom is not 'a-tomic', indivisible. At the turn of the century, J. J. Thomson's investigation of cathode rays was demonstrating that atoms contain electrons, which they will eject under special circumstances. Thus atoms became divisible, and a structure could be visualized in which each is composed of negatively charged electrons, one or many, and a body of some sort bearing a compensating positive charge.

Thomson's work left unclear whether its positive charge is smeared throughout an atom, forming a pudding in which the electrons are embedded, or is concentrated at some localized places in the atom. Ten years later Ernest Rutherford's experiments showed that all the positive charge in an atom is indeed very strictly localized in one nucleus, which also carries most of the mass of the atom. Thus an atom could be visualized as a solar system writ small—a collection of tiny electrons whirling about a tiny nucleus much as the planets circulate about the sun. In such a picture an atom consists mostly of empty space: it is far from being the tiny infrangible ball, everywhere dense, that was probably visualized by Dalton.

As a principal tool in his investigations, Rutherford used the so-called alpha particles ejected by the radioactivity of radium. Studies of such radioactivity initiated the intensive work of our century, showing that the atomic nuclei in turn are divisible and have a structure.

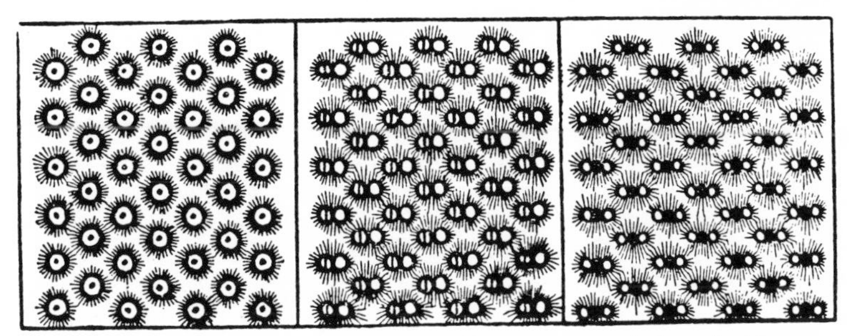

FIG. 1.3. At the beginning of the nineteenth century, John Dalton proposed that the particles of any form of matter consist of combinations of atoms of various elementary substances in simple numerical ratios. He pictured his idea in this way.

What remains of Democritus' contention that the divisibility of matter must come to an end, of Dalton's assertion that 'the existence of ultimate particles of matter can scarcely be doubted'? The atoms on the surface of the filament in a vacuum tube emit electrons: the atoms are not strictly 'atomic'. Slowly in a luminous watch dial, catastrophically fast in an atomic bomb, substances change their species: the elements are not strictly 'elementary'. But atoms can still be dissolved and recaptured; they bounce off the inner walls of automobile tyres; they can be formed into 'atomic beams' and deflected. And we can still repeat Dalton's statement that 'chemical analysis and synthesis go no further than to the separation of particles one from another, and to their reunion'.

The reconciliation of atomic divisibility with atomic integrity, and of emptiness with solidity, remains a philosophical problem. But a picture that successfully describes the results of experiments can perhaps be taken as a partial reconciliation, and such a picture has been constructed in this century. Its construction has required revolutionary modifications of the mechanical principles that have served so well for objects larger than a single atom. Quantum theory, developed by Max Planck in a different context while Thomson was studying cathode rays, was applied in 1913 to Rutherford's planetary model of the atom by Niels Bohr, to provide the beginnings of the needed reconciliation. The wave-mechanical formulation of that theory, invented by Erwin Schrödinger in 1925, has been especially successful in providing an orderly way of describing atomic behaviour, and this monograph is largely occupied with that form of description.

These ways of thinking leave little of the concreteness that ordinary experience makes comfortable to us. The 'tiny particles' disappear into objects whose behaviour is describable only by wave equations, recalling the vision of certain early mystics that matter is 'frozen light'. To visualize events in the subatomic world, we must reconstitute our ways of visualization, perhaps even accepting the limitation that we can do no better than provide means for calculating consistently whatever adds up to affect our senses and our instruments.

2. The Chemical Elements

ROUGHLY speaking, a million single definite substances are known today. In accordance with the programme first visualized by Robert Boyle in 1661 they can all be described as combinations of ninety elementary materials. And the elaboration of the ideas proposed by John Dalton in 1808 has produced for most of these substances very concrete pictures, in which they appear made of tiny molecules, each consisting of relatively few atoms of their elementary ingredients.

The identification of what substances are elementary and what are not captured much attention over two hundred years, and the table of naturally occurring elements was only completed during the first quarter of this century (Fig. 2.1). In our own time the table of elements is being lengthened by the construction of 'artificial' elements. Deliberate bombardment of natural elements by particles of very high energy has welded those particles to the bombarded nuclei, producing tiny quantities of elements not otherwise encountered and usually short-lived. But the substances that make the stable world remain accountable in terms of ninety elements.

And of those elements, no more than twenty suffice to account for most of the world. Living matter in all its forms consists almost entirely of carbon, hydrogen, oxygen, and nitrogen. Smaller quantities of other elements are needed for life, especially calcium, sulphur, and phosphorus in bones and proteins, and magnesium, iron, sodium, potassium, chlorine, and iodine elsewhere. The inorganic crust of the earth consists mostly of fluorides, oxides, silicates (silicon and oxygen), and carbonates (carbon and oxygen) of aluminium, magnesium, calcium, iron, sodium, and potassium. The earth's core is believed to be made largely of molten iron and nickel. Thus the rich diversity of the world flows not from the diversity of its elementary parts but from the versatility of about twenty atomic species in their ways of joining together.

The naturally occurring elements are listed in Table 2.1. Most are metals. There are fourteen familiar nonmetals: hydrogen, oxygen, nitrogen, carbon, fluorine, chlorine, bromine, iodine, sulphur, selenium, phosphorus, arsenic, boron, and silicon. Six more, less familiar, are the *rare gases*: helium, neon, argon, krypton, xenon, and radon, which are unique in their reluctance to enter into chemical combination. Three elements, tellurium, antimony, and germanium, fall on the borderline between metals and nonmetals, sharing some properties of both classes.

Long before this listing was completed, chemists noticed more refined

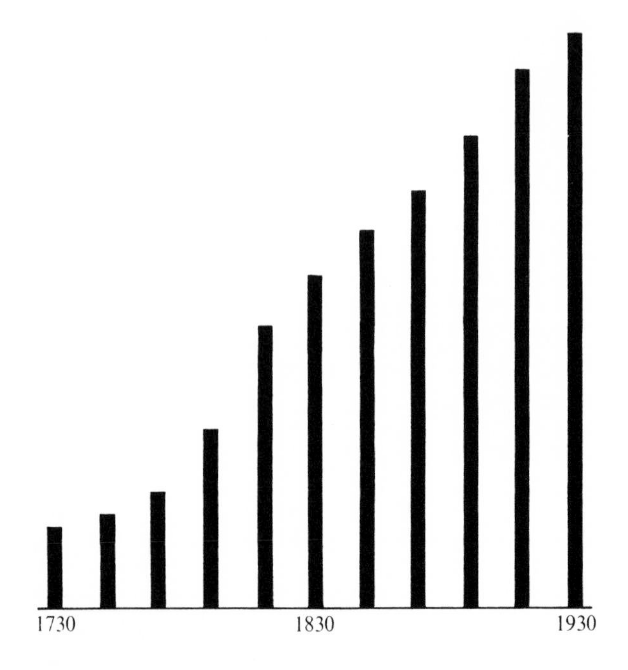

FIG. 2.1. The number of identified chemical elements grew over two centuries from twelve in 1730 to ninety in 1930.

similarities and differences between the elements, which led them to subclassify the metals and the nonmetals. Beginning in 1817, Johann Döbereiner called attention to certain triads of elements with similar properties, and pointed out that the members of each triad can be listed in an order such that the middle member has properties representing a mean between the two extremes. One such triad is chlorine, bromine, and iodine, belonging to a group that we now call *halogens*. Another is lithium, sodium, and potassium, today called *alkali metals*. A third, the *alkaline earth metals*, is calcium, strontium, and barium.

From 1830 to 1860, many chemists extended this kind of correlation, and showed that such a relationship is not necessarily confined within a triad. They added fluorine to the halogens, magnesium to the alkaline earths; and when rubidium and caesium were discovered in 1860–1, they joined the family of the alkalis. And two other families received recognition, oxygen, sulphur, selenium, and tellurium forming one, and nitrogen, phosphorus, arsenic, antimony, and bismuth the other.

It might seem more reasonable on physical grounds to place the gases oxygen and nitrogen in the same family, and so also the two metalloids tellurium and bismuth. But the compelling grounds for relating oxygen to tellurium and nitrogen to bismuth were chemical, and their explanation in physical terms in our own century exemplifies the close fraternity of chemistry and physics in elucidating the structure of matter. These families were founded in the chemical ideas of *combining weight* and *valency*, described in Discussion 2.1, which arose from Dalton's work at the beginning of the

Discussion 2.1

COMBINING WEIGHT, ATOMIC WEIGHT, AND VALENCY

The *combining weight* of an element is that weight that combines with, or displaces from combination, a unit weight of hydrogen. For example, eight grams of oxygen combine with one gram of hydrogen to form water. Thus the combining weight of oxygen is 8.

According to the atomic hypothesis, the elements combine to form molecules of which each contains whole numbers of its constituent elementary atoms. The simplest immediate assumption is that a molecule of water might be written HO, with one atom of hydrogen and one of oxygen. Then, if the *atomic weight* of hydrogen is taken as unity, the atomic weight of oxygen would be 8, in agreement with its combining weight.

But the volumes of gaseous hydrogen and oxygen that combine to form water are as two to one; and when water is decomposed by electrolysis, twice as large a volume of hydrogen as of oxygen is evolved. Early in the nineteenth century, chemists concluded that equal volumes of two gases (at the same temperature and pressure) contain equal numbers of molecules, regardless of the species of the gases. Then if the molecules of gaseous hydrogen and oxygen are composed of the same number of atoms, water must be written H_2O, not HO, and the atomic weight of oxygen must be 16, not 8.

The number of combining weights in one atomic weight of an element is called the *valency* of the element, and accordingly oxygen receives valency two. By extending such reasoning in various ways to the experimentally measured combining weights of other elements, an internally consistent set of atomic weights and valencies has been assigned to them.

nineteenth century and rapidly grew in definiteness and application. The elements that form the family containing oxygen and tellurium all have in common the valency two, and the family containing nitrogen and bismuth shows the valency three.

The observation of such families spurred many efforts to find a systematic way of organizing the entire list of elements that would reflect their chemical similarities and differences. These attempts began to meet with success in the 1860's; by that time about sixty-five elements had been identified, their combining weights were known, and valencies had been assigned to most of them. Several chemists noticed a periodic type of regularity appearing when the elements are listed in order of increasing atomic weight. John Newlands, in particular, observed that often the properties of an element are similar to those of the element seven places further along in the list of the elements known in his time, and enunciated a 'law of octaves' in analogy with the characteristics of the diatonic musical scale.†

Finally, in 1869, Lothar Meyer in Germany and Dmitri Mendeleev in Russia independently put forward essentially similar versions of what is now called the *periodic table of the elements.* Mendeleev's form of the table emphasized especially clearly the periodic variation of the valencies of the elements. By listing the elements in horizontal rows in order of increasing atomic weight, and beginning a new row when the valency repeats, the vertical columns of the table constitute the families of related elements that had been noticed earlier, which are now called *groups.*

Only a wide experience with descriptive chemistry can give an appreciation of the large body of isolated observations set in order by the periodic table. It is now the central intellectual tool of chemistry. Some of its great power appears in the predictions that Mendeleev himself made with its aid. He left gaps in the table when the ascending sequence of atomic weights revealed gaps in the sequence of valencies, and predicted confidently that new elements would be discovered that would fill those gaps. And he predicted in some detail what properties those elements would turn out to have. Table 2.2 shows, for example, his predictions in 1871 regarding germanium (which he called eka-silicon, Es) and their verification by Clemens Winkler fifteen years later. With the same success he predicted the existence and properties of scandium and gallium.

Mendeleev's earliest grouping by valency provided seven groups, strongly suggesting Newlands' octaves. In 1871, he revised the form of the table to accommodate certain heavier elements in an eighth group. Then, beginning in 1894, the chemically inert rare gases were successively discovered, and Mendeleev suggested that they be accomodated in a ninth group, with valency zero. But the atomic weights of the lighter of these gases, especially of neon, filled gaps in Mendeleev's eighth group, and they were placed there even though they exhibited no family resemblance to such elements as iron and platinum already placed in that group. And a few other anomalies persisted. The rare gas argon, whose atomic weight is greater than that of potassium, must precede potassium on grounds of valency; and the sequence of tellurium and iodine is anomalous in a similar fashion.‡

Two large successions of physical discovery in this century have rationalized the periodic table in a way that removes these anomalies. The first began with the identification of the electron by J. J. Thomson in 1897, leading to the picture of an atom as containing electrons whose number, running through the successive integers, exactly parallels the sequence in the periodic table. The second

† Reading his paper to a meeting of the Chemical Society of London in 1865, Newlands met with some scorn; in particular the chemist Carey Foster asked him whether he had thought to classify the elements in order of the initial letters of their names.

‡ Such anomalies led Sir William Ramsay, codiscoverer with Lord Rayleigh of most of the rare gases, to describe the study of the periodic table as 'a painful pleasure'.

began with Ernest Rutherford's discovery in 1913 that the mass of an atom is principally lodged in a nucleus at the atom's core, leading to a picture of nuclei as bearing a positive charge equal and opposite to that of the atom's electrons. Henry Moseley soon made it clear that the true sequence of elements in the periodic table is that of their *atomic number*, the number of electrons that they contain, rather than their atomic weight.

It may seem odd that the atomic weight, due mostly to the nucleus, correlates so closely with the atomic number: the weight increases as roughly twice the number. As early as 1815, William Prout had noticed that the atomic weights of many of the lighter elements are simple multiples of the atomic weight of hydrogen, and suggested that hydrogen might be the sole primary source of all matter, from which the other elements are produced by some process of condensation. As more atomic weights were determined accurately, 'Prout's hypothesis' had to be discarded.

But that hypothesis is an interesting precursor of a picture of nuclei that is especially useful to chemists today, one that has removed the remaining anomalies of the periodic table. In this picture a nucleus is thought to be made of *protons*, the particle with the mass and charge of the nucleus of hydrogen, and *neutrons*, the particle, identified by James Chadwick in 1932, with the same mass as the proton but without charge. Then the atomic number of a nucleus is equal to its number of protons, and the atomic mass is equal to that of the sum of its protons and neutrons. The departure of the atomic weight of a naturally occurring element from a simple integer can be ascribed mostly† to the fact that the element contains a mixture of atoms having the same atomic number and different atomic masses. Indeed by the technique of mass spectroscopy, such *isotopes* can now be separated and identified. The natural distribution of the isotopes of an element is remarkably uniform throughout the earth's crust, giving significance to a precise chemical determination of atomic weight.

A study of the various forms of the periodic table that are shown in Tables 2.3, 2.4, and 2.5 will repay those who are not already well acquainted with this way of relating the properties of the elements. The way in which the atomic pictures due to Rutherford and Niels Bohr rationalized the periodic table is described in the following chapter, and Chapter 8 provides a wave-mechanical discussion of the table.

† Tiny residual departures are now understood also. The mass of a neutron differs slightly from that of a proton; and the *binding energy* of the nuclear constituents makes a small contribution to the atomic mass, as Einstein's famous relation $E = mc^2$ predicts.

3. Models of an Atom

DURING the later half of the nineteenth century, while chemists were identifying the elementary species of atoms and developing the periodic table to summarize their growing knowledge of how these species behave, more strictly physical investigations of the electrical properties of matter were gradually making clear that atoms are not indivisible, but are aggregates of separate particles. Studies of the rays emitted from the cathode (the negatively charged electrode) in an evacuated tube carrying an electrical discharge showed that these 'cathode rays' consist of tiny particles bearing a negative charge. Since cathodes made of any sort of metal emit the particles, J. J. Thomson concluded that they are ingredients of all species of atoms. And he showed that these particles, which received the name *electrons*, are all alike, each with a mass only about $\frac{1}{2000}$ as large as that of a hydrogen atom.

In 1904, Thomson suggested that the arrangement of the electrons within atoms might provide a structural origin for the relations between the elements correlated in the periodic table. He guessed that the electrons might be found in groups or layers, and that the number of electrons in the outermost layer might largely determine the chemical properties of the species (Fig. 3.1). The rare gas atoms might contain especially stable arrangements of electrons. An atom with one electron less than a rare gas atom, for example chlorine, might tend to acquire an extra electron and so form a negative ion. An atom with one more electron, for example sodium, might readily lose it, to form a positive ion. Hence atoms that readily lose electrons would combine chemically with atoms that readily acquire electrons.

Ernest Rutherford's discovery in 1912, that the positive charge in an atom is concentrated in a single nucleus at its core, suggested the exciting possibility that all these properties of atoms might be explained by the mechanical behaviour of their constituent particles. It was natural to assume that each electron in an atom traverses an orbit about the nucleus, whose size and shape is determined by balancing the inward attractive force between the negative charge on the electron and the positive charge on the nucleus against the outward centrifugal force of the circulating electron. The attractive force, obeying Coulomb's law, would vary inversely as the square of the distance between the charges, in other words, by the same rule that governs the gravitational attraction between a planet and the sun.

But clearly such a picture would require some modification in order to explain two conspicuous properties of atoms. The first is the fact that an atom of a particular species is indistinguishable from any other atom of the

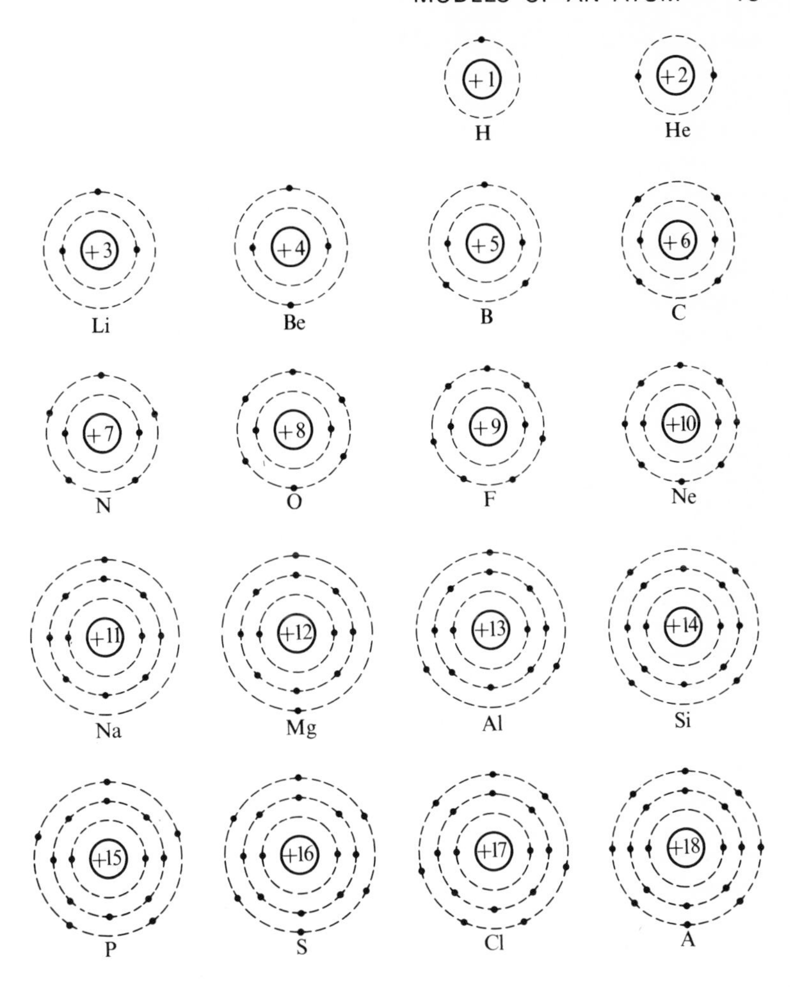

FIG. 3.1. J. J. Thomson's idea that the electrons in an atom occupy *shells* is still useful for discussing the chemical behaviour of the atomic species, especially those in the first three periods of the periodic table. In each successively heavier species of atom, one electron is added and one additional positive charge is added also, along with the increased mass. In atoms of the rare gases all shells that are occupied at all are filled to their maximum capacity.

same species under the same conditions anywhere in the world at any time. Since atoms are constantly colliding with one another, it is remarkable that they remain unchanged and characteristic of their species. Solar systems that started out alike would not remain so after such vicissitudes. For example, a collection of solar systems could not be condensed from a gas to a liquid and revaporized without change.

In the second place it can be shown that, if an electron is a charged particle traversing an orbit, it should radiate electromagnetic waves, such as light waves. Since these waves would carry energy away from the atom, the electron should lose energy, and its orbit should change, allowing it to come constantly closer to the nucleus. As its orbit changes, the frequency with which it traverses the orbit would also change, and thus the frequency of the wave that it radiates should change. Finally the electron should lose enough energy to fall into the nucleus; the atom should collapse to a size not much bigger than its nucleus.

But atoms are not continually radiating and collapsing. They radiate energy in this way only when energy has been imparted to them. And then the frequencies of their radiations are very sharply defined and characteristic of each atomic species—so characteristic that breaking up the radiation into its component colours in a spectroscope is used analytically for identifying the species.

The year after Rutherford's discovery, Niels Bohr made the first attempt to provide a mechanical explanation for these pecularities of the atomic planetary system. He recognized that the *quantum theory*, developed by Max Planck in 1900, could be applied to the behaviour of the electrons in atoms. It offered a reason why an electron might adopt a state of motion from which it could be dislodged only with difficulty, a state chosen from a limited set of such states. And Bohr recognized also that, when it is forced to change from one of its permitted states of motion to another, the electron might emit a burst of radiation whose frequency would be sharply defined by its initial and final states.

Planck had invented the quantum theory to explain the results of experiments on the light radiated from a hole in an empty black cavity, heated until it glowed. Presumably that light is given off by the atoms at the walls of the cavity: their heat vibrations rapidly oscillate the electric charges associated with them, and cause them to radiate the electromagnetic waves called light, much as a radio antenna radiates similar waves with longer wavelengths.

Examining the different amounts of energy radiated at different wavelengths by such a *black body*, Planck was forced to conclude that the vibrating atoms do not emit or absorb light continuously but only in little packages. Each package has a definite size, in terms of its content of energy, which is directly proportional to the frequency of the light. In other words, the energy in each package is given by $E = h\nu$, where ν is the frequency, and h is the constant of proportionality now called *Planck's constant*.

In 1905, Albert Einstein made several applications of these new ideas to other puzzling phenomena, for example, the photoelectric effect, in which certain metals emit electrons when they are irradiated with light of suitable frequencies. In discussing such effects, he suggested that each package of radiant energy, $h\nu$, did not spread out from its source in all directions but was sent out unidirectionally, like a particle. It has become customary to speak of such a 'particle' as a *photon.*

Einstein's success in explaining these matters gave the quantum theory further stature. Bizarre though the theory seemed, it yielded answers that checked the results of experiments that no theory had been able to explain before. Bohr began by applying that theory to the planetary model of hydrogen, confining attention to the possible circular orbits of the single electron in the atom and requiring that the electron's angular momentum be *quantized*, or, in other words, take only certain definite values.

It is remarkable that his consideration of this simple example enabled Bohr to formulate two far-ranging postulates that remain valid today, despite the radical revision of the quantum theory that began in 1925. Bohr's first postulate asserts that an atomic system can exist in certain *stationary states* without radiating its energy. Each state corresponds to a definite value of the total energy E of the system. A transition of the system from one stationary state to another is attended by a gain or loss of energy, in an amount equal to the difference between the energies of the two states. That energy difference may be emitted or absorbed as radiation, or transmitted to or from some other atomic system. Bohr's second postulate relates to the frequency of the radiation emitted or absorbed by the system when it makes a transition between two states. If the energies of the two states are E_2 and E_1, the frequency of the radiation is given by $h\nu = E_2 - E_1$.

These postulates provide a programme for calculating from experimental data the energies of the presumed stationary states of the electron in a hydrogen atom. First, measure in a spectrometer the frequencies of all the radiations that hydrogen emits when it is excited. Second, convert the frequencies to energy differences by using Bohr's rule. Third, find what energies E form a consistent set that yield these differences (Fig. 3.2).

The success of Bohr's calculations (Discussion 3.1) for his model of hydrogen spurred others to develop generalizations of his theory, suited to a wider variety of mechanical systems. But when these methods were applied to atomic systems only slightly more complicated than the hydrogen atom, they were unsatisfactory. They could not be made to yield the energy levels spectroscopically determined in helium, the atom with two electrons instead of just one. Nor did they give correct answers for the *hydrogen-molecule ion*—the hydrogen molecule from which one electron has been removed and the one remaining electron circulates about two nuclei. And they offered no convincing explanation of why the electrons in a many-electron atom should adopt configurations displaying the *shell* structure that had been proposed by Thomson to explain the behaviour summarized in the periodic table.

Chemists pursued Thomson's proposal nevertheless, assuming that a satisfactory physical justification for the idea of shell structures would eventually support its chemical plausibility. Formal electronic pictures were

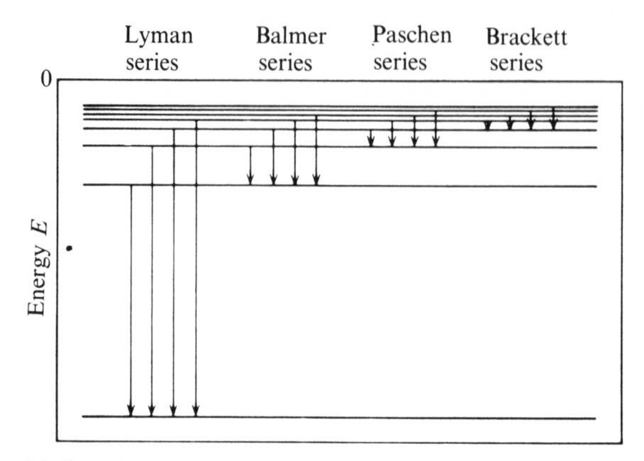

FIG. 3.2. Some levels of energy in the hydrogen atom, worked out from the observed spectral frequencies by using Bohr's frequency postulate, $h\nu = E_2 - E_1$. The various series of spectral frequencies are named after the spectroscopists who examined them. The Lyman series lies in the ultraviolet frequency range. Some frequencies of the Balmer series lie in the visible range. The Paschen and Brackett series lie in the infrared range.

constructed from 1916 onwards, especially by W. Kossel, G. N. Lewis, and I. Langmuir, setting in order the chemical behaviour of the atomic species. Kossel paid especial attention to the circumstances in which one atom would transfer electrons to another, making each atom an ion with the electronic configuration of a rare gas atom. Lewis showed that chemical bonds not ascribable to the transfer of electrons, for example, the bond between the two hydrogen atoms in a hydrogen molecule, can be explained by supposing that the atoms acquire the electronic configurations of rare gas atoms by sharing electrons.†

In 1924, Louis de Broglie opened an entrance to the path that physicists and chemists have followed since in developing the mechanics of electrons in atoms and molecules. He suggested that a train of *waves* might be essentially associated with any particle. His idea reciprocated Albert Einstein's earlier suggestion that light, previously thought of as waves, behaves in some respects like a stream of particles, the photons. Taken together, the two suggestions imply a curious symmetry in nature, by which both wavelike and particlelike behaviour can be discerned in all of nature's fundamental processes.

Three years later, Clinton Davisson and Lester Germer confirmed de Broglie's idea by experiments in which electrons were diffracted by a crystal of nickel; all agreed that diffraction is a phenomenon to be expected of waves, and not of 'particles' as they had previously been understood. And already in 1926, without waiting for such confirmation of de Broglie's idea, Erwin Schrödinger had moved ahead to develop a *wave mechanics* that treats all

† The bonds formed between atoms in chemical compounds are discussed in detail in *Bonds between atoms*.

Discussion 3.1

QUANTIZING CIRCULAR ORBITS

For his first model of the hydrogen atom in 1913, Bohr assumed that the one electron moves at a speed v in a circle of radius r centred on the nucleus. Then Coulomb's law gives the attractive force between the nucleus and the electron as $e^2/4\pi\epsilon_0 r^2$, where e is the magnitude of the electronic charge and the nuclear charge alike. Since this force is the centripetal force that keeps the electron on the circle, given by mv^2/r, where m is the mass of the electron, the radius of the circle is related to the velocity of the electron by

$$\frac{mv^2}{r} = \frac{e^2}{4\pi\epsilon_0 r^2}. \tag{D3.1}$$

On such an electronic motion, Bohr imposed the quantum condition that the angular momentum of the electron about the nucleus should be able to adopt only those values that are integral multiples of $h/2\pi$, where h is Planck's constant.

$$mvr = n\frac{h}{2\pi}, \tag{D3.2}$$

where n is an integer. Then the permitted radii are obtained by eliminating the velocity between (D3.1) and (D3.2):

$$r = \frac{\epsilon_0 n^2 h^2}{\pi e^2 m}. \tag{D3.3}$$

In other words, the radii of the permitted orbits vary as n^2. The smallest permitted orbit, obtained by setting $n = 1$ and using the values $e = 1{\cdot}6 \times 10^{-19}$ C, $m = 9{\cdot}035 \times 10^{-31}$ kg, and $h = 6{\cdot}55 \times 10^{-34}$ J s, turns out to have a radius $a = 5{\cdot}28 \times 10^{-11}$ m.

In order to compare this model of the hydrogen atom with experimental measurements on hydrogen, the energies of the electron in the various permitted orbits must be calculated, and their differences must be compared with the frequencies of observed spectral lines by using Bohr's second postulate

$$E_2 - E_1 = h\nu. \tag{D3.4}$$

The potential energy of the electron at a distance r from the nucleus, relative to a zero of energy for an electron infinitely distant, is

$$U_{\text{pot}} = \frac{-e^2}{4\pi\epsilon_0 r}. \tag{D3.5}$$

In the expression for the kinetic energy of the electron, $\frac{1}{2}mv^2$, (D3.1) can be used to eliminate the velocity, and hence

$$U_{\text{kin}} = \frac{e^2}{8\pi\epsilon_0 r}. \tag{D3.6}$$

Thus the total energy of the electron is obtained with the assistance of (D3.3) as

$$E = U_{\text{kin}} + U_{\text{pot}} = -\frac{e^2}{8\pi\epsilon_0 r} = -\frac{e^4 m}{8h^2 n^2 \epsilon_0^2}. \tag{D3.7}$$

Hence, on passing from an outer orbit with energy E_2 to an inner orbit with energy E_1, the electron should emit radiation with the frequency given by (D3.4) as

$$h\nu = \frac{e^4 m}{8h^2\epsilon_0^2}\left(\frac{1}{n_1^2} - \frac{1}{n_2^2}\right). \tag{D3.8}$$

Spectral lines are ordinarily measured in terms of the wavelength λ rather than the frequency of the radiation, and the data are converted to *wave numbers*, $\tilde{\nu} = 1/\lambda$. Since $\lambda\nu = c$, where $c = 3 \times 10^8$ m/s is the velocity of light, (D3.8) is best converted for comparison with experiment into

$$\tilde{\nu} = R_{\text{H}}\left(\frac{1}{n_1^2} - \frac{1}{n_2^2}\right), \tag{D3.9}$$

where $R_{\text{H}} = e^4 m/8\epsilon_0^2 h^3 c$ is known as the Rydberg constant. The startling success of Bohr's model lay in the fact that (1) the observed lines of the hydrogen spectrum could indeed be indexed by such numbers as n_1 and n_2 in (D3.9), and (2) the calculated value, $R_{\text{H}} = 10\,967\,776\ \text{m}^{-1}$, agreed with the observed value, $R_{\text{H}} = 10\,967\,818\ \text{m}^{-1}$, to the sixth significant figure.

particles as if they were waves. He used it to explain the same spectroscopic data from hydrogen that had supported Bohr's earlier model, and wrote, 'The new conception can be generalized, and I believe that it penetrates deeply into the true nature of the quantum rules'.

Since then, wave mechanics has been rapidly extended, explaining successfully such matters as the bonding of atoms in molecules and solids, and the behaviour of electrons in them. This newer mechanics pictures an atom as a positive nucleus surrounded by a continuous distribution of negative electricity with a variable density, contributed by all the electrons. The stationary states† permitted to the electrons are mathematically analogous to the standing waves of vibration permitted in a stretched string held at both ends, or in a vibrating drumhead.

The classical picture and the newer picture of an electron in an atom can be compared most readily by looking at the simplest imaginable case of a confined particle. Imagine a particle free to move along a line but confined within a definite segment of it. More graphically portrayed, the particle finds itself within a tube of narrow bore, capped at both ends. In the classical picture, the particle rushes back and forth along the line, reversing the sign of its momentum p each time it is reflected at an end of the line.

De Broglie's proposal pictures the particle by a wave, of wavelength λ, and connects that wavelength with the momentum by the relation

$$p = h/\lambda, \tag{3.1}$$

where h is Planck's constant. Since the particle is rigorously confined to the line segment, of length L, its wave must vanish outside that segment, and it must fall smoothly to zero at each end of the segment. The wave is of sinusoidal form, and hence its wavelength can be no greater than twice the length of the segment, as Fig. 3.3 suggests.

The wavelength can be shorter than that; but in order to fulfill the same requirements, it can take only certain definite values. It must be an integral submultiple of the maximum wavelength (Fig. 3.4). Thus the relation (3.1) proposed by de Broglie restricts the momentum of the particle to definite values corresponding with those possible values of the wavelength.

If the particle is subject to no forces within the line segment to which it is confined, its potential energy can be taken as zero everywhere within that segment. Since its energy is then entirely kinetic, and equal to $p^2/2m$, where m is its mass, its permitted levels of energy can be written at once as

$$E_n = \frac{h^2n^2}{8mL}. \tag{3.2}$$

Notice two unexpected consequences of this way of picturing a confined particle. In the first place, the particle cannot be at rest. Because its maximum wavelength is finite, its minimum momentum is not zero, according to (3.1).

† A monograph in this series, *Stationary states*, summarizes those aspects of wave-mechanics that are used in the remainder of this monograph.

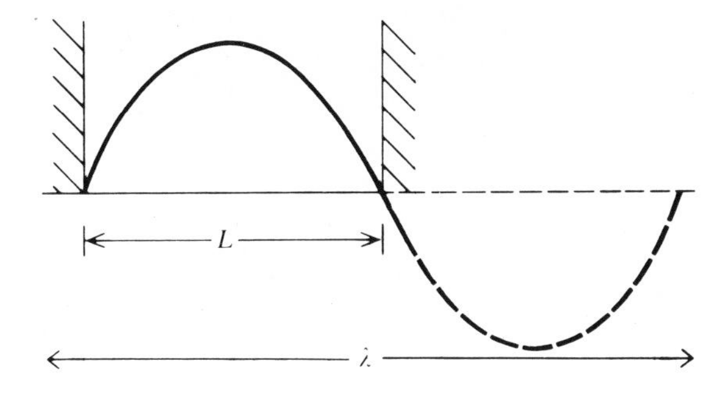

FIG. 3.3. The de Broglie wavelength λ of a particle, restricted to a line segment of length L and moving freely within that restriction, cannot be greater than $2L$.

Indeed, the more closely it is confined, the faster it must move and the greater its kinetic energy must be. In the second place, its energy can have only certain well-defined values and cannot take intermediate values, according to (3.2).

Both these consequences are more general than this example might suggest. In particular, the permitted energies, associated with permitted orbits by the earlier atomic theory, become associated with permitted waves in the newer. As Schrödinger wrote in his first paper on wave mechanics, 'The introduction of integers arises in the same natural way as, for example, in a vibrating string, for which the number of nodes is integral'.

The particle whose behaviour has just been examined is often called 'the particle in a box'. It is convenient to think of this and of any other confined particle as moving in a *potential well*, whose shape is visualized by plotting the potential energy of the particle as a function of its position x. For the

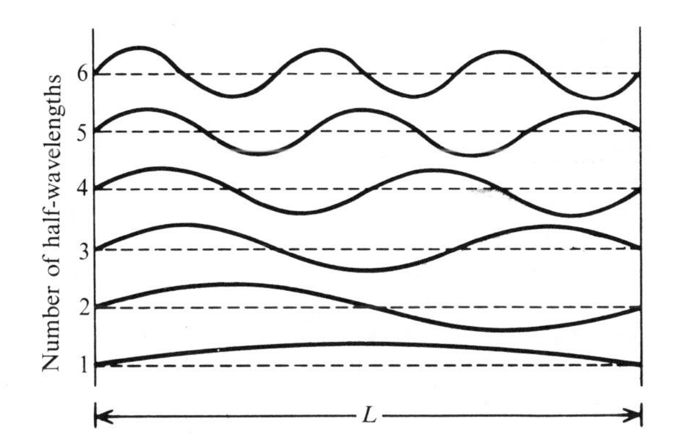

FIG. 3.4. The permitted waves for a particle restricted to a line segment L have wavelengths given by $n\lambda/2 = L$, where n is an integer.

particle in a box, the potential energy is zero everywhere inside the well and infinite everywhere outside it. The energies permitted to the particle can be plotted at appropriate levels in the well (Fig. 3.5).

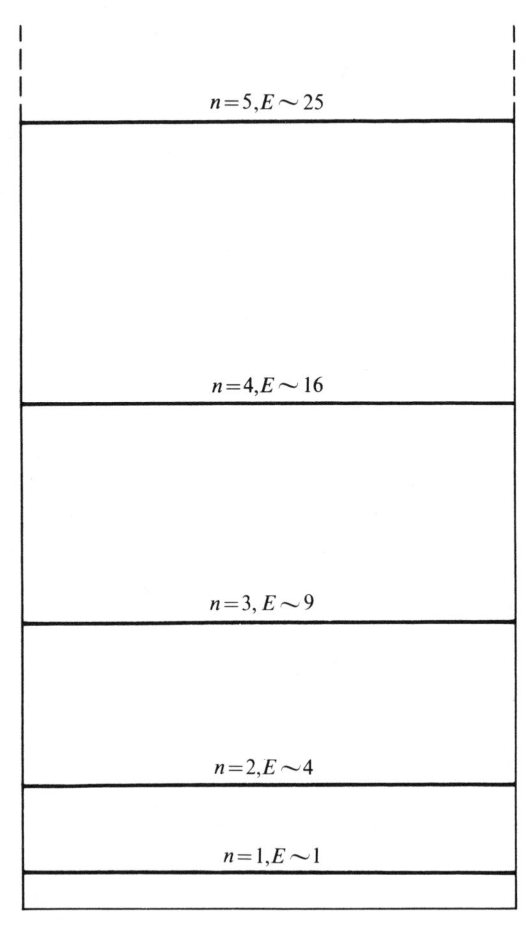

FIG. 3.5. The energy of a particle restricted to the line segment of Fig. 3.4 can take only certain values proportional to n^2, where n is an integer.

In shifting attention to a particle in a potential well with some other shape, the determination of the waveforms and the associated energies is complicated by the fact that the momentum of the particle, and thus its wavelength, will vary as it proceeds across the well. The wavelength is connected with the kinetic energy, by way of the relations $U_{\text{kin}} = p^2/2m$ and $p = h/\lambda$; but it is the total energy E, and not U_{kin}, that is conserved in the motion.

Discussion 3.2

A PARTICLE WITH VARYING WAVELENGTH

When the potential energy of a particle varies with its position, its momentum will vary also, and thus its *wavelength* will not be uniquely defined. A way to find the right forms of the waves for the particle in such a case can be suggested by the following argument. But the fact that this way will turn out to be correct cannot be proved by such an argument.

Examine first the sinusoidal waveforms of the simple waves with well-defined wavelengths shown in Fig. 3.4. Each is described mathematically by

$$\psi = A \sin \frac{2\pi}{\lambda} x. \tag{D3.10}$$

Differentiating (D3.10) twice with respect to x yields

$$\frac{d^2\psi}{dx^2} = -A\frac{4\pi^2}{\lambda^2} \sin \frac{2\pi}{\lambda} x. \tag{D3.11}$$

Thus the wave form ψ has the property

$$\frac{d^2\psi}{dx^2} = -\frac{4\pi^2}{\lambda^2}\psi. \tag{D3.12}$$

Now assume that ψ will possess this property even when λ varies with x. In order to examine the results of that assumption, λ can be expressed in terms of the way in which the potential energy varies with x, by using de Broglie's relation (3.1) in the following way.

The total energy $E = U_{kin} + V$, where V is the potential energy. Hence

$$U_{kin} = \frac{p^2}{2m} = E - V, \tag{D3.13}$$

and thus by (3.1)

$$\lambda = \frac{h}{\sqrt{\{2m(E-V)\}}}. \tag{D3.14}$$

Substituting the expression (D3.14) for λ in (D3.12) finally yields

$$\frac{d^2\psi}{dx^2} + \frac{8\pi^2 m}{h^2}(E-V)\psi = 0. \tag{D3.15}$$

This is Schrödinger's equation for describing the wave shape of a particle moving in one dimension. The particular wave shapes sought are those solutions to the equation that are physically sensible. When the particle is bound to a potential well, the only sensible solutions are those that approach zero as the particle departs increasingly from the well. Invariably such solutions can be found only for certain definite values of E. Thus there arise discrete permitted *states* for the particle, each with an associated value of the total energy.

A suitable way for describing a particle whose wavelength varies with its position is suggested in Discussion 3.2. The discussion develops a differential equation whose solutions specify appropriate wave shapes. As in the case of the particle in a box, the only acceptable solutions to the equation are those which approach zero in regions where the potential energy of the particle increases without limit. Again it turns out that those solutions correspond with a set of distinct values of the total energy E.

A familiar example of a confined particle is the *simple harmonic oscillator*: a particle attracted toward a fixed point by a force that increases in direct proportion to its distance from that point. When the origin of the coordinate x is chosen at the fixed point, and the potential energy of the particle is taken

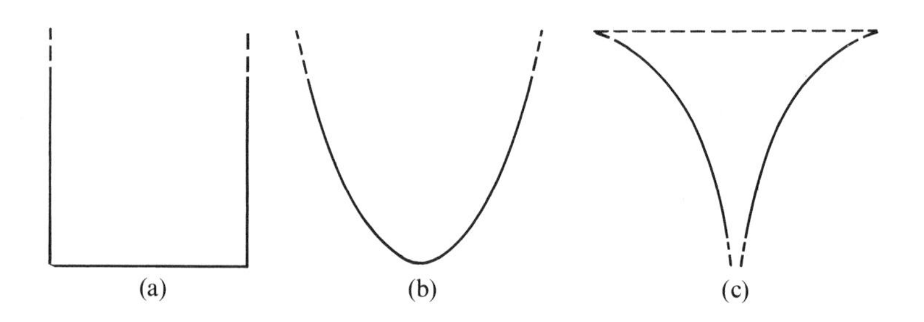

FIG. 3.6. Potential wells for (a) a particle in a box and (b) a harmonic oscillator; and (c) a cross-section of the well for the electron in a hydrogen atom.

to be zero when it is at that point, the restoring force can be described by $F = -kx$, and the potential energy by $V = \frac{1}{2}kx^2$, where k is a constant that measures the 'stiffness' of the oscillator. In other words, the well has a parabolic shape. Then it can be shown† that the permitted energies are evenly spaced along the energy scale, increasing in increments of $h\nu$, where ν is the frequency of the oscillator, and that the lowest permitted energy is $\frac{1}{2}h\nu$.

Turning finally to the electron in a hydrogen atom, you find that the appropriate well, described by (D3.5), has three features that are qualitatively different from the foregoing wells. First, this well has no natural bottom, but descends indefinitely, narrowing as it goes. Second, the well has a natural top, which now provides the natural zero of potential energy. And third, the well has more dimensions because it is appropriate to a particle moving in three dimensions instead of one. By displaying a cross-section of it, as if the electron were constrained to move along a single line through the nucleus, Fig. 3.6 contrasts the atomic well with the wells for the box and for the harmonic oscillator.

The behaviour of an electron moving in such a well is analysed exactly

† The simple harmonic oscillator is discussed quantitatively in *Stationary states*, Chapter 5.

in Chapters 5, 6, and 7. Again there is a lowest permitted level of energy, corresponding with a wave shape called the *ground state*, and the other permitted energies form a discrete set. With increasing energy, however, the permitted levels pack closer together on the energy scale. And when the electron has an energy greater than zero, all energies are permitted to it; it has escaped from the well and can cruise through space unhindered, and the waves that describe it no longer approach zero as it departs from the well. The distributions of energy levels characteristic of the three wells of Fig. 3.6 are contrasted in Fig. 3.7.

Notice that the wave-mechanical model of the hydrogen atom possesses the quantitative property that marked the greatest achievement of Bohr's planetary model (Discussion 3.1); the permitted energies are spaced as

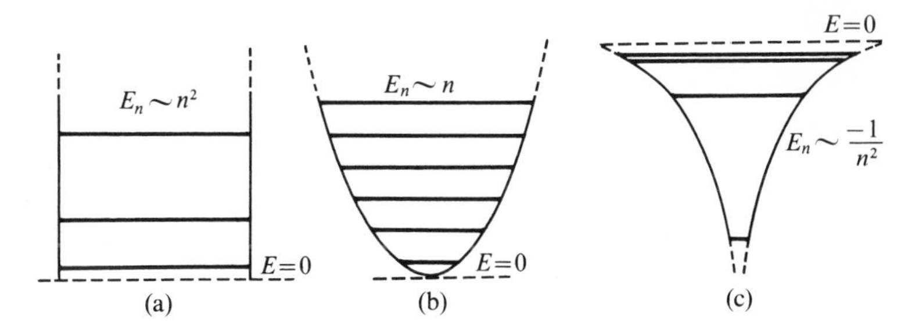

FIG. 3.7. Some permitted energy levels E_n for (a) a particle in a box, (b) a harmonic oscillator, and (c) the electron in a hydrogen atom.

$-1/n^2$, where n is an integer. Moreover, the wave-mechanical method can be extended successfully to the more complicated cases for which the planetary method failed. But the successful extension of any model to explain the electronic shell structures of atoms with more than two electrons must invoke two further properties of the electron, discovered in 1925.

The first is the fact that an electron behaves as if it were a spinning ball; it exhibits an intrinsic angular momentum about an axis through its centre. Furthermore, that angular momentum appears to be very rigorously quantized. In the first place, it has a fixed absolute value, the same for all electrons. In the second place, whenever a system contains more than one electron, the spin of each is either in the same direction or in the opposite direction to the spins of its companions. Hence the description of any electron, begun by specifying its wave, must be completed by specifying whether its spin is 'up' or 'down'.

The second property is another unexpected limitation placed on each electron by the presence of companion electrons. Apparently no two electrons can have the same components of spin, and the same form of wave in the same part of space, at once. If two electrons have the same wave, one must direct its spin 'up' and the other 'down', and no third electron can have the same wave. In consequence of this *exclusion principle*, each of the possible waves in an atom can be ascribed to no more than two electrons. Since the possible shapes of waves are spatially distributed in a somewhat shell-like way, the property described by the exclusion principle forces additional electrons to acquire waves that fall in shells corresponding to successively higher energies. In this way the periodic table can be understood in much detail, as Chapter 8 describes.

The remainder of this monograph is devoted to examining more exactly the wave-mechanical picture of an atom. For an introduction to the exact study, the next chapter discusses some examples of the waves for particles moving in one dimension in a well that, like the atomic well, has a top.

4. A Particle in a One-dimensional Well

THE last chapter portrayed an atom as consisting of electrons in a *well*. The well is constructed for the electrons by the attraction between their negative charges and the positive charge on the atom's nucleus. Any atomic nucleus is at least 2000 times heavier than an electron, and one can study the behaviour of the electrons quite successfully by taking the mass of the nucleus to be infinite.

In beginning a more exact study of the behaviour of the electrons in an atom, it is helpful to examine the expected behaviour of electrons in stationary wells, which, though rather artificial, are simpler than atomic wells. Much can be learned, for example, by neglecting the three-dimensionality of an atomic well, and concentrating first on the fact that such a well has a top.

The simplest one-dimensional well that has a top can be constructed by chopping off, at some finite point, the infinitely high top of the more familiar 'box' shown in the last chapter. To be sure, the resulting well differs from that for the atom by having a bottom. But only such a bottom will keep the electron from falling indefinitely down the well, so long as the walls of the well are a finite distance apart. Fig. 4.1 shows the suggested well, of depth $|V|$ and width l, with its top rather than its bottom now placed at the zero of energy.

Before diving into the formalities of this problem, pause to look at the physics of what you are doing. If the box had infinitely high walls, an electron inside it would be obliged to have at least the kinetic energy of the ground state of such a box. When the height of the walls is reduced, you might expect that there would come a point at which the kinetic energy exceeds the loss of potential energy that the electron experiences while it is in the box. In other words, the total energy of the electron would become positive, and large enough to carry it over the top of the box, and it would no longer be confined.

From these arguments you might expect, but you cannot be sure, to find boxes so narrow and shallow as to offer the electron no bound states at all. As a matter of fact, it turns out that there is at least one bound state for any one-dimensional box. There need not be more than one; and in a three-dimensional box there need not be even one. But there can at best be only a finite number of such states, for their energies must be less than that of the top of the box, and greater than that of the bottom of the box; and since the top of the box does not flare outward, the permitted energies cannot pack closer together near the top as they can in an atom.

It is instructive to work in detail the problem of the box shown in Fig. 4.1, for some of the ideas that emerge are useful in other problems. Schrödinger's

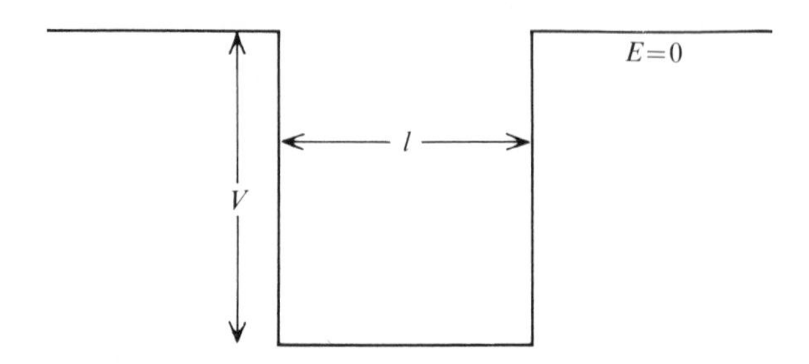

FIG. 4.1. The one-dimensional box with walls of finite height.

equation (see Discussion 3.2) for the stationary states is

$$\frac{h^2}{8\pi^2 m}\frac{d^2\psi}{dx^2}+[E-V(x)]\psi = 0, \tag{4.1}$$

where E is the electronic energy associated with the electronic state ψ, $V(x)$ is the potential energy of the electron when its coordinate has the value x, m is its mass, and h is Planck's constant. Here $V(x)$ is not an analytic function of x at the boundaries of the box, but everywhere else it is the simplest sort of function: zero outside the box, and V (a negative constant) inside. Hence we can seek a solution to the equation that applies outside the box, and a solution *for the same value of* E to the equation inside the box, and match the two solutions properly at the boundaries of the box.

Eqn (4.1) and the consequences that flow from it can be written more simply if the factor $h^2/8\pi^2 m$ is removed at the outset by changing the scale of some of the quantities in the equation. The usual way to do this in atomic problems is to express energies and distances in the *atomic units* described in Discussion 4.1. The equation then reads

$$\frac{d^2\psi}{dx^2}+[E-V(x)]\psi = 0. \tag{4.2}$$

For the wave functions ψ_o outside the well, the equation is

$$\frac{d^2\psi_o}{dx^2}+E\psi_o = 0, \tag{4.3}$$

and for the wave functions ψ_i inside the well, the equation is

$$\frac{d^2\psi_i}{dx^2}+(E-V)\psi_i = 0. \tag{4.4}$$

The bound states of the electron are those whose values of E are less than zero, or below the top of the well. Hence E is negative in (4.3) for the states of interest, and the most general useful solution is

$$\psi_o = A_+ \exp\sqrt{(-E)}x+A_- \exp -\sqrt{(-E)}x, \tag{4.5}$$

Discussion 4.1

ATOMIC UNITS

When Schrödinger's equation is written for electrons, it is customary to simplify the appearance of the equation by using *atomic units* of length and energy. The atomic unit of length is

$$a_0 = \frac{h^2\epsilon_0}{\pi me^2},$$

and of energy is

$$W_H = \frac{me^4}{8h^2\epsilon_0^2},$$

where h is Planck's constant and e and m are the charge and mass of an electron. Then, starting with Schrödinger's equation in ordinary units for a particle moving in one dimension,

$$\frac{h^2}{8\pi^2 m}\frac{d^2\psi}{dx^2}+[E-V(x)]\psi = 0,$$

and making the substitution

$$E \to \frac{me^4}{8h^2\epsilon_0^2}E, \qquad V \to \frac{me^4}{8h^2\epsilon_0^2}V,$$

$$x \to \frac{h^2\epsilon_0}{\pi me^2}x,$$

converts the equation to

$$\frac{d^2\psi}{dx^2}+[E-V(x)]\psi = 0.$$

It is noteworthy that the atomic unit of energy (often called the Rydberg) is the negative of the energy of an electron in the ground state of a hydrogen atom, $-13{\cdot}53$ electron volts, and the atomic unit of length is equal to the radius of the corresponding *Bohr orbit*, $0{\cdot}53 \times 10^{-10}$ m (see Discussion 3.1).

where A_+ and A_- are arbitrary constants. But these functions must be so chosen that they do not increase without limit as the electron goes infinitely far from the well. Hence the wave functions to the left of the well must be

$$\psi_l = A_l \exp \sqrt{(-E)}x, \tag{4.6}$$

and those to the right of the well must be

$$\psi_r = A_r \exp -\sqrt{(-E)}x. \tag{4.7}$$

Inside the well, V is negative, and E is less negative because E is above the bottom of the well. Since $E-V$ is positive, the general solution to (4.4) is

$$\psi_i = A_s \sin \sqrt{(E-V)}x + A_c \cos \sqrt{(E-V)}x. \tag{4.8}$$

Previous work† with symmetrical wells leads you to expect that the wave functions for this case will be either symmetric or antisymmetric about the centre of the well. Then, if the zero of x is chosen at the centre of the well, those two sorts of solutions should take the forms:

$$\begin{aligned}
\text{Sym:} \quad & \psi_l = A_o \exp \sqrt{(-E)}x, \\
& \psi_r = A_o \exp -\sqrt{(-E)}x, \\
& \psi_i = A_i \cos \sqrt{(E-V)}x; \\
\text{Anti:} \quad & \psi_l = A_o \exp \sqrt{(-E)}x, \\
& \psi_r = -A_o \exp -\sqrt{(-E)}x, \\
& \psi_i = A_i \sin \sqrt{(E-V)}x.
\end{aligned} \tag{4.9}$$

Turn now to matching the various parts of these functions at the places where they must join:

ψ_l to ψ_i at $x = -l/2$, and ψ_r to ψ_i at $x = +l/2$. By (4.9) those requirements imply

$$\begin{aligned}
\text{for Sym:} \quad & A_o \exp -\sqrt{(-E)}l/2 = A_i \cos \sqrt{(E-V)}l/2; \\
\text{for Anti:} \quad & A_o \exp -\sqrt{(-E)}l/2 = -A_i \sin \sqrt{(E-V)}l/2.
\end{aligned} \tag{4.10}$$

But there is another requirement: the slopes of the functions must match at these points, as well as the functions themselves. This requirement emerges from looking at what would happen if the slopes did *not* match. If at some point the slope of a function has a discontinuity, then at that point the rate of change of the slope is infinite. The rate of change of the slope of ψ is $d^2\psi/dx^2$, and (4.3) and (4.4) say that $d^2\psi/dx^2$ cannot be infinite so long as E, V, and ψ are finite. When (4.9) is differentiated, the additional requirement can be written:

$$\begin{aligned}
\text{for Sym:} \quad & A_o\sqrt{(-E)}\exp -\sqrt{(-E)}l/2 = A_i\sqrt{(E-V)}\sin \sqrt{(E-V)}l/2; \\
\text{for Anti:} \quad & A_o\sqrt{(-E)}\exp -\sqrt{(-E)}l/2 = A_i\sqrt{(E-V)}\cos \sqrt{(E-V)}l/2.
\end{aligned} \tag{4.11}$$

Eliminating A_o and A_i between (4.10) and (4.11) yields the equations that determine the permitted energy levels:

$$\begin{aligned}
\text{for Sym:} \quad & \tan\sqrt{(E-V)}l/2 = \sqrt{\left(\frac{E}{V-E}\right)}; \\
\text{for Anti:} \quad & \tan\sqrt{(E-V)}l/2 = -\sqrt{\left(\frac{V-E}{E}\right)}.
\end{aligned} \tag{4.12}$$

These are transcendental equations. For any given values of V and l, which specify the box, the values of E that satisfy these equations are the

† For example, the particle in a box and the harmonic oscillator, discussed in *Stationary states*, Chapters 3 and 5.

permitted energy levels. The easiest way to see what these roots will be like is to express E as a fractional multiple of V, the depth of the well: $E = \epsilon V$. Notice that ϵ is positive when E is negative (as in bound states) and V is negative (as inside the well). Then (4.12) becomes:

$$\text{for Sym:} \qquad \tan \frac{l\sqrt{(-V)}}{2}\sqrt{(1-\epsilon)} = \sqrt{\left(\frac{\epsilon}{1-\epsilon}\right)};$$

$$\text{for Anti:} \qquad \tan \frac{l\sqrt{(-V)}}{2}\sqrt{(1-\epsilon)} = -\sqrt{\left(\frac{1-\epsilon}{\epsilon}\right)}. \tag{4.13}$$

The factor $l\sqrt{(-V)}/2$ now specifies the well, and ϵ can run from 0 to 1. Fig. 4.2 shows what the right sides of the equations look like, and Fig. 4.3 shows what $\sqrt{(1-\epsilon)}$ looks like.

If $l\sqrt{(-V)}$ is less than π, the left sides of the equations never become negative, and thus there is no antisymmetric state. There will always be a symmetric state, however, for with increasing ϵ the left side decreases monotonically from $\tan l\sqrt{(-V)}/2$ to zero while the right side of the first of (4.13) increases monotonically from zero to infinity; and there must be a value $\epsilon = \epsilon_0$ for which the two sides are equal (Fig. 4.4). The smaller the value of $l\sqrt{(-V)}$, the smaller ϵ_0 will be; in other words, as the well gets narrower or shallower, the energy of the bound state gets nearer the top of the well, but there is always a bound state. Fig. 4.5 shows roughly the shape of the wave function.

When $l\sqrt{(-V)}$ is greater than π, an antisymmetric solution becomes possible; when $l\sqrt{(-V)}$ is greater than 2π, a second symmetric solution appears; and so on. At the same time, the values of ϵ corresponding to the solutions first found approach nearer and nearer to unity. In order to follow what happens as this process continues, take $l\sqrt{(-V)} = N\pi$ where N is a very large number. Then there will be a root of (4.13), alternately symmetric and antisymmetric, somewhere in each of the ranges $\sqrt{(1-\epsilon)} = 0\text{–}1/N$, $1/N\text{–}2/N$, $2/N\text{–}3/N$, and so on. Taking the upper bound of each range as approximating a value of $\sqrt{(1-\epsilon)}$ that yields a root, and remembering that $1-\epsilon$ measures the fractional distance of an energy level above the bottom of the well, you see that the sequence $1-\epsilon = 1/N^2, 4/N^2, 9/N^2 \ldots n^2/N^2$ corresponds to the distribution of energy levels for the box with infinitely high walls. Here, however, the sequence breaks off at $n = N$.

Starting with a well in which $l\sqrt{(-V)}$ is large and then reducing it, you can think of the progressive disappearance of the higher-energy states as a process of squeezing them out of the well (if you reduce l) or pushing them up out of the well (if you reduce V). There is an especially simple sort of well that can be constructed by a limiting process in which all but one state is squeezed out by letting l go to zero, and that one state is preserved by letting V become negatively infinite, in such a way that the product Vl stays constant. That well is infinitely narrow and infinitely deep: the *delta well*.

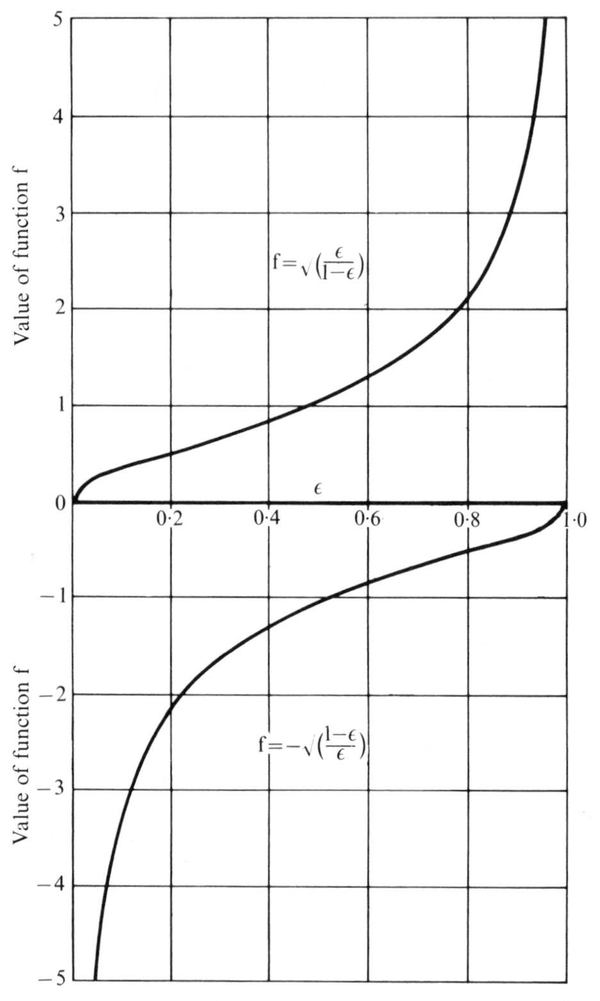

FIG. 4.2. Plots of $\sqrt{(\epsilon/1-\epsilon)}$ and $-\sqrt{(1-\epsilon/\epsilon)}$, to assist in inferring the roots of (4.13).

Fig. 4.5 and eqn (4.2) provide the means for guessing the form of the state that the delta well preserves. The wave function will fall off exponentially from the well in both directions. The space inside the well, within which the wave function of Fig. 4.5 'turns around', is now reduced to nothing, and the wave function must turn around infinitely rapidly at the location of the well, exhibiting a 'corner' there. But that is consistent with eqn (4.2), which says that since $V(x)$ is negatively infinite at the well, the rate of change of the slope

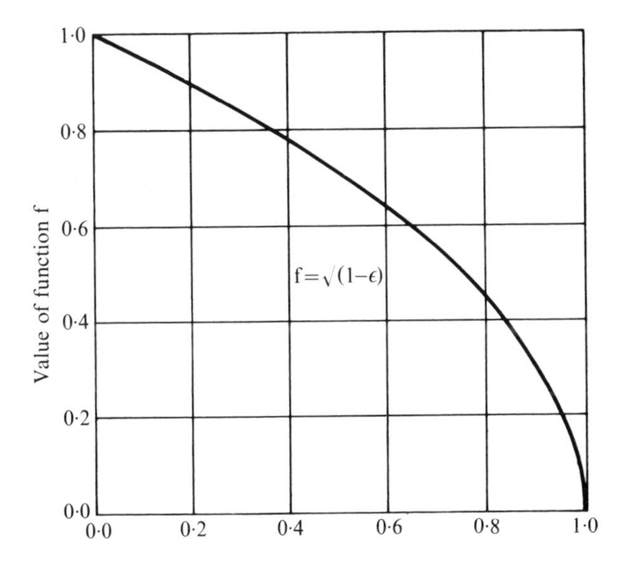

FIG. 4.3. Plot of $\sqrt{(1-\epsilon)}$, for the significant range of ϵ.

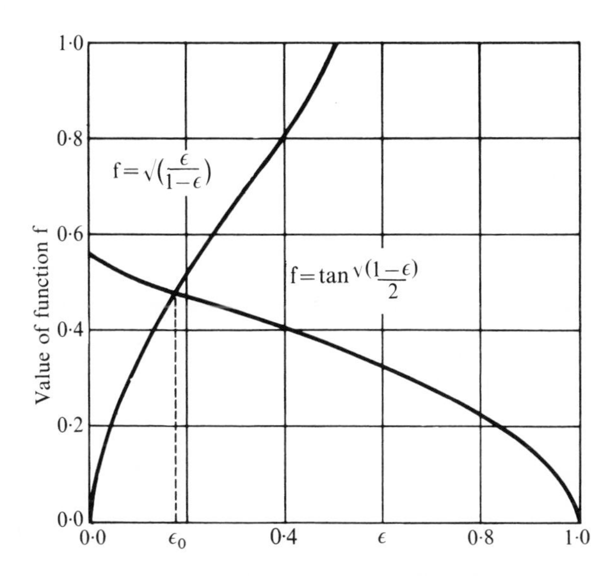

FIG. 4.4. The only root of (4.13) when $l\sqrt{(-V)} = 1$.

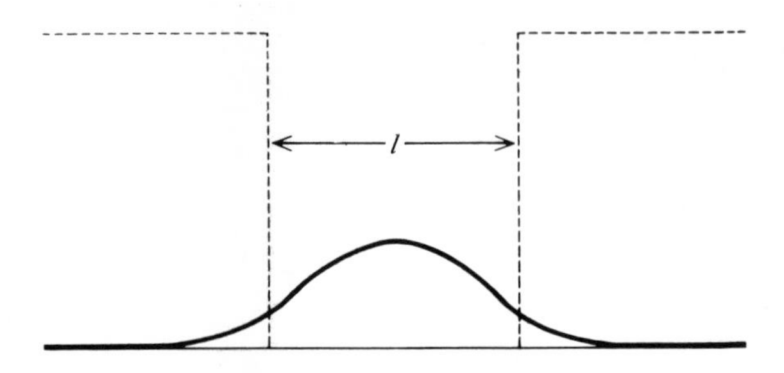

FIG. 4.5. The wave function of the particle in its ground state in the one-dimensional box with finite walls.

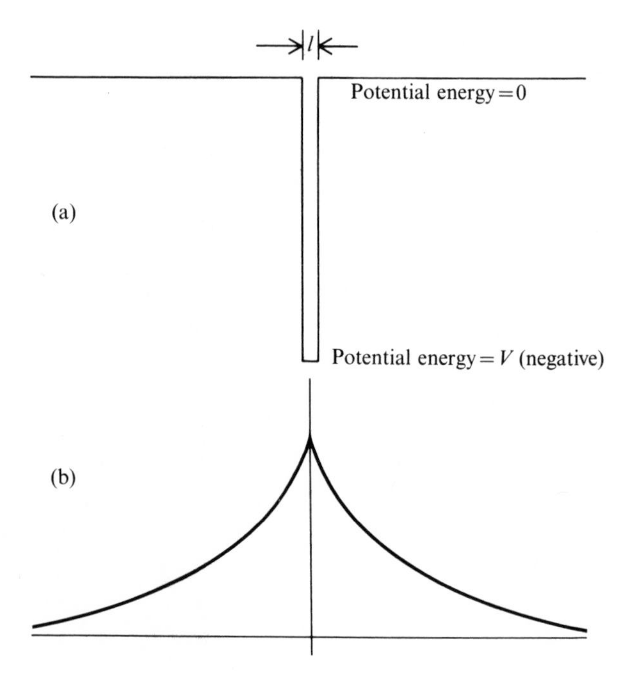

FIG. 4.6. The wave function of the particle moving in one dimension in the only bound state that the one-dimensional delta well affords.

Discussion 4.2

THE DELTA WELL

The wave function for the state of a particle in the presence of a delta well can be found in the following way. Since the entire function is outside the well, (4.3) applies everywhere except at the point $x = 0$, and the solutions (4.6) and (4.7) are appropriate for the two sides of the well. Since they must match at the well, $A_l = A_r$. It remains to determine what the relation between the slopes on the two sides is—what the angle at the corner is.

That relation comes from integrating (4.2) across a well that is taken to be very narrow and very deep, and then proceeding to the limit. At the well V is so large (negatively) that E is negligible in (4.2); hence the equation can be multiplied by $\mathrm{d}x$ and written

$$\mathrm{d}\frac{\mathrm{d}\psi}{\mathrm{d}x} = V\psi \,\mathrm{d}x. \tag{D4.1}$$

The width l of the well is so small that ψ retains its value for $x = 0$ over that small distance. Integrating (D4.1) over l yields

$$\left.\frac{\mathrm{d}\psi_r}{\mathrm{d}x}\right|_{x=0} - \left.\frac{\mathrm{d}\psi_l}{\mathrm{d}x}\right|_{x=0} = \psi(0)\,.\,Vl, \tag{D4.2}$$

which becomes

$$-2\sqrt{(-E)} = Vl. \tag{D4.3}$$

Thus E acquires a finite value if l approaches zero and V approaches minus infinity in such a way that the product is finite. It is convenient to denote that product Vl by minus the positive constant η, so that

$$\sqrt{(-E)} = \frac{\eta}{2}. \tag{D4.4}$$

Thus a single parameter η specifies the well, and also through (D4.4) the energy of the bound state associated with it and the wave function

$$\begin{aligned} \psi_l &= A\exp(\eta/2)x, \\ \psi_r &= A\exp(-\eta/2)x. \end{aligned} \tag{D4.5}$$

is infinite, unless the wave function vanishes there. The wave function will look like Fig. 4.6; its mathematical form is derived in Discussion 4.2.

You will find the delta well a convenient rough model for an atom in a later discussion of the two-electron atom (Chapter 9). It can also give insight in discussing the chemical bond and the behaviour of electrons in solids. This model strips the atom down to a minimum: a one-dimensional well that has a top and no bottom. The energy in the one bound state that the well affords can be adjusted to correspond with the energy in any particular state of any particular atom.

PROBLEMS

4.1 Determine the value of the constant A in (D4.5) that normalizes the wave function for the bound state associated with a delta well.

4.2 Describe the justification for the three steps in the following calculation of the expectation value of the potential energy of an electron in the bound state associated with a delta well, where A is the normalizing constant:

$$U_{\text{pot}} = \int_{-\infty}^{\infty} \psi^* V \psi \, \mathrm{d}x = A^2 \int V \, \mathrm{d}x \quad \text{(integrated over the distance } l\text{)}$$

$$= -A^2 \eta.$$

4.3 Using the results of Problems 1 and 2, calculate the expectation value of the *kinetic* energy of an electron in the bound state associated with a delta well.

5. A Particle in a Central Field

IN an atom, an electron is actually in a three-dimensional well, not a one-dimensional well such as the last chapter discussed. The most important new feature that three-dimensionality introduces is the possibility that the electron can *revolve in an orbit*, acquiring angular momentum about the nucleus. The uncertainty principle of quantum mechanics places limits on how much can be known about that angular momentum.† In the case of a particle in a central field, the principle allows you to specify only two aspects of the angular momentum independently and simultaneously: the absolute value of the total angular momentum, and the value of the component of angular momentum about one chosen axis.

Now, of course, an isolated atom offers no special axis to choose, unless an applied magnetic or electric field or the like establishes a special direction in space. Nor does an atom in a 'cubically symmetric' environment offer a unique axis: the atom cannot distinguish between the cubic axes x, y, and z. In the first place, then, examine what can be learned about the angular dependence of the wave functions for the states of an electron in an atom without introducing a preferred axis, postponing consideration of its angular momentum to the next chapter.

The examination begins with an atom containing only one electron. This may seem a poverty-stricken choice: most atoms contain more than one electron. Even a hydrogen atom is normally accompanied by another, forming together the diatomic hydrogen molecule, in which two electrons and two nuclei interact. But one-electron atoms do occur, and comprise the only type for which exact results can be derived. The description of other atoms uses insights and approximations that flow from those results. Some appropriate ways of using these insights will be described in Chapters 8 and 10.

When only one electron is present, the nucleus of the atom presents the electron with a potential well that is spherically symmetrical:

$$V = -Ze^2/4\pi\epsilon_0 r,$$

where r is the distance of the electron from a nucleus of atomic number Z and charge Ze. The discussion in this chapter confines its attention to the fact that V is a function only of r, without concern for the exact form of that function. In other words the derived results are limited only by the fact that the field in which the electron moves is a *central field*.

† The uncertainty principle, and its application to angular momenta, are discussed in *Stationary states*, Chapters 3 and 8.

Schrödinger's equation for the stationary states of the electron in a central field, written in atomic units (Discussion 4.1), reads

$$\frac{\partial^2\psi}{\partial x^2}+\frac{\partial^2\psi}{\partial y^2}+\frac{\partial^2\psi}{\partial z^2}+[E-V(r)]\psi = 0, \tag{5.1}$$

where $r = \sqrt{(x^2+y^2+z^2)}$. The method usually used for finding the desired solutions to this equation begins by transforming from the coordinates (x, y, z) to spherical polar coordinates (r, θ, ϕ) centred on the nucleus. But, as Fig. 5.1 shows, the transformation to (r, θ, ϕ) subtly introduces a unique

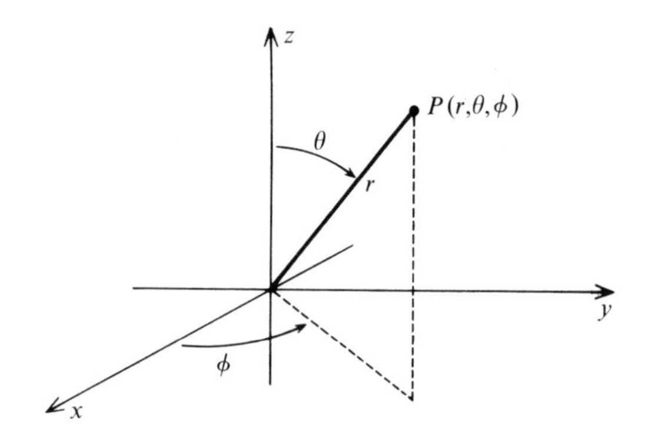

FIG. 5.1. The transformation of coordinates from (x, y, z) to (r, θ, ϕ) gives the z direction a unique position.

axis, the polar axis from which the angle of colatitude θ is measured. It is worthwhile to pursue a method somewhat less usual, which retains the Cartesian axes, so that the resulting expressions are functions of x, y, and z, and are appropriate to a choice of three equivalent mutually perpendicular axes rather than one unique axis.

In either case the method of solution proceeds by *separation of variables.* Using spherical polar coordinates, you would seek solutions of the form $R\Theta\Phi$, where R is a function of r alone, Θ of θ alone, and Φ of ϕ alone. Using Cartesian coordinates, you seek solutions of the form RL, where R is again a function of $r = \sqrt{(x^2+y^2+z^2)}$ alone and L is a function of direction alone. Directions can always be specified by *direction cosines.* Fig. 5.2 shows that in Cartesian coordinates the direction cosines of the direction from the origin of coordinates to the point (x, y, z) are x/r, y/r, z/r. Hence L will be a function of x/r, y/r, and z/r. Notice that x/r, y/r, and z/r are homogeneous functions of x, y, and z, of degree zero, and thus that L must also be such a

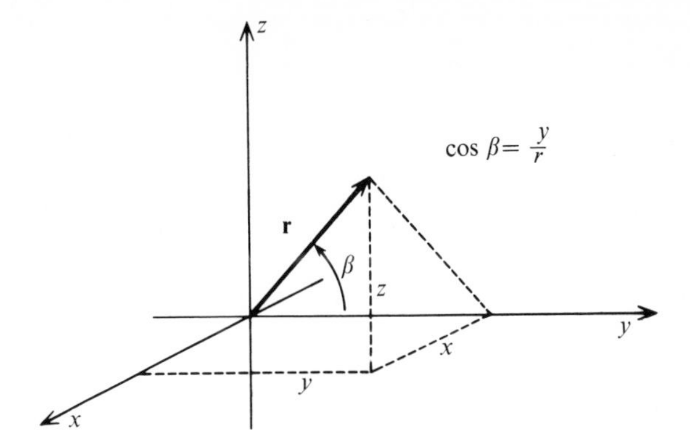

FIG. 5.2. The direction cosines of a vector r relative to the axes are $\cos\alpha = x/r$, $\cos\beta = y/r$, $\cos\gamma = z/r$.

function. The substitution $\psi = RL$ in (5.1) yields

$$L\left[\frac{\partial^2 R}{\partial x^2}+\frac{\partial^2 R}{\partial y^2}+\frac{\partial^2 R}{\partial z^2}\right]+R\left[\frac{\partial^2 L}{\partial x^2}+\frac{\partial^2 L}{\partial y^2}+\frac{\partial^2 L}{\partial z^2}\right]+$$
$$+2\left[\frac{\partial R}{\partial x}\frac{\partial L}{\partial x}+\frac{\partial R}{\partial y}\frac{\partial L}{\partial y}+\frac{\partial R}{\partial z}\frac{\partial L}{\partial z}\right]+$$
$$+[E-V(r)]RL = 0. \quad (5.2)$$

First examine the third bracket, containing the products of first derivatives. Since R is a function of r alone, $\partial R/\partial x = \mathrm{d}R/\mathrm{d}r\;\partial r/\partial x$, etc. But differentiation of $r = \sqrt{(x^2+y^2+z^2)}$ yields $\partial r/\partial x = x/r$, etc., and hence the bracket becomes

$$\frac{2}{r}\left(x\frac{\partial L}{\partial x}+y\frac{\partial L}{\partial y}+z\frac{\partial L}{\partial z}\right)\frac{\mathrm{d}R}{\mathrm{d}r}.$$

It is useful to apply to this bracket Euler's theorem for homogeneous functions, derived in Discussion 5.1. According to that theorem, if F is a homogeneous function of x, y, and z of degree n, then $x\,\partial F/\partial x+y\,\partial F/\partial y+z\,\partial F/\partial z = nF$. Since L is a homogeneous function of degree zero, $n = 0$, and this bracket vanishes.

Look next at the first bracket in (5.2). A second differentiation like that of the last paragraph gives

$$\frac{\partial^2 R}{\partial x^2} = \frac{\mathrm{d}^2 R}{\mathrm{d}r^2}\left(\frac{\partial r}{\partial x}\right)^2+\frac{\mathrm{d}R}{\mathrm{d}r}\frac{\partial^2 r}{\partial x^2}$$
$$= \frac{x^2}{r^2}\frac{\mathrm{d}^2 R}{\mathrm{d}r^2}+\left(\frac{1}{r}-\frac{x^2}{r^3}\right)\frac{\mathrm{d}R}{\mathrm{d}r}, \text{ etc.} \quad (5.3)$$

Discussion 5.1

EULER'S THEOREM FOR HOMOGENEOUS FUNCTIONS

A function of several variables is *homogeneous of order n* if the result of multiplying each of the variables by a constant λ is the same as multiplying the function by a power n of the constant:

$$f(\lambda x, \lambda y, \lambda z) = \lambda^n f(x, y, z).$$

Now temporarily regard λ as a *variable*, and differentiate that defining expression partially with respect to λ. This operation produces:

on the left,

$$\frac{\partial f}{\partial(\lambda x)}\frac{\partial(\lambda x)}{\partial\lambda}+\frac{\partial f}{\partial(\lambda y)}\frac{\partial(\lambda y)}{\partial\lambda}+\frac{\partial f}{\partial(\lambda z)}\frac{\partial(\lambda z)}{\partial\lambda} = x\frac{\partial f}{\partial(\lambda x)}+y\frac{\partial f}{\partial(\lambda y)}+z\frac{\partial f}{\partial(\lambda z)};$$

on the right,

$$n\lambda^{n-1}f.$$

Then, setting $\lambda = 1$, you obtain Euler's result,

$$x\frac{\partial f}{\partial x}+y\frac{\partial f}{\partial y}+z\frac{\partial f}{\partial z} = nf.$$

Hence the first bracket in (5.2) becomes $L[\mathrm{d}^2R/\mathrm{d}r^2+(2/r)(\mathrm{d}R/\mathrm{d}r)]$. When the second bracket is transposed to the right side, and the equation is divided by RL, it can be rewritten

$$\frac{1}{R}\left(\frac{\mathrm{d}^2R}{\mathrm{d}r^2}+\frac{2}{r}\frac{\mathrm{d}R}{\mathrm{d}r}\right)+E-V(r) = -\frac{1}{L}\left(\frac{\partial^2L}{\partial x^2}+\frac{\partial^2L}{\partial y^2}+\frac{\partial^2L}{\partial z^2}\right). \qquad (5.4)$$

At this point it is tempting to paraphrase the argument usual in a separation of variables, saying 'the left side of (5.4) is a function of r alone, and the right side is a function of direction alone; each must therefore be equal to a constant k, independent of both r and direction'. But that argument would not be valid here: L is a function of direction alone but its derivatives are taken with respect to coordinates, not directions, and thus are not functions of direction alone.

The suggested argument therefore requires a slight modification. Every time a homogeneous function is differentiated with respect to one of its variables, a new homogeneous function arises of an order one less than that of the original function. Hence the right side of (5.4) must be a homogeneous function of order minus two. Now the only homogeneous function of that sort that is a function of r alone, and thus can equal the left side of (5.4), is k/r^2, where k is a constant. Hence each side of (5.4) can be set equal to k/r^2, and the separated equations can be written:

$$\frac{d^2R}{dr^2}+\frac{2}{r}\frac{dR}{dr}+\left[E-\frac{k}{r^2}-V(r)\right]R = 0; \tag{5.5}$$

$$\frac{\partial^2 L}{\partial x^2}+\frac{\partial^2 L}{\partial y^2}+\frac{\partial^2 L}{\partial z^2}+\frac{k}{r^2}L = 0. \tag{5.6}$$

Solutions to (5.5) will give the radial dependence of the wave functions. Solutions to (5.6) will give the directional dependence, which is of primary interest at the moment.

The desired solutions for L must be rational integral functions of the direction cosines, and therefore polynomials in x/r, y/r, and z/r. This fact suggests seeking solutions L_l that are homogeneous polynomials of degree l in those direction cosines. Any such polynomial has the form $L_l = H_l/r^l$, where H_l is a homogeneous polynomial of degree l in x, y, and z. After this form is substituted into (5.6), it reads

$$\frac{1}{r^l}\left(\frac{\partial^2 H_l}{\partial x^2}+\frac{\partial^2 H_l}{\partial y^2}+\frac{\partial^2 H_l}{\partial z^2}\right)-\frac{2l}{r^{l+2}}\left(x\frac{\partial H_l}{\partial x}+y\frac{\partial H_l}{\partial y}+z\frac{\partial H_l}{\partial z}\right)+$$
$$+[l(l-1)+k]\frac{H_l}{r^{l+2}} = 0. \tag{5.7}$$

Euler's theorem for homogeneous functions, applied to the second bracket in (5.7), yields

$$x\frac{\partial H_l}{\partial x}+y\frac{\partial H_l}{\partial y}+z\frac{\partial H_l}{\partial z} = lH_l. \tag{5.8}$$

Then, multiplying by r^{l+2} and transposing the term in H_l, you can rewrite (5.7):

$$r^2\left(\frac{\partial^2 H_l}{\partial x^2}+\frac{\partial^2 H_l}{\partial y^2}+\frac{\partial^2 H_l}{\partial z^2}\right) = [l(l+1)-k]H_l. \tag{5.9}$$

The left side of (5.9) has the factor $r^2 = x^2+y^2+z^2$, and therefore the right side must either vanish or have the factor r^2 also. But if it has the factor r^2, then $L_l = r^2P/r^l = P/r^{l-2}$, and corresponds to a case of L_{l-2} in which H_{l-2} does *not* have the factor r^2. You will therefore miss no solutions by supposing that the right side of (5.9) vanishes: it is permissible to take $k = l(l+1)$ and

$$\frac{\partial^2 H_l}{\partial x^2}+\frac{\partial^2 H_l}{\partial y^2}+\frac{\partial^2 H_l}{\partial z^2} = 0. \tag{5.10}$$

In other words, the H_l will be the homogeneous polynomials of degree l that satisfy Laplace's equation, often called the *spherical harmonics*; and

Discussion 5.2

POLYNOMIALS SATISFYING LAPLACE'S EQUATION

The most general homogeneous polynomial of degree l in three variables can be written in the form

$$H_l = a_0 z^l + z^{l-1}(a_{11}x + a_{10}y) + z^{l-2}(a_{22}x^2 + a_{21}xy + a_{20}y^2) + \ldots$$
$$+ (a_{ll}x^l + a_{l,l-1}x^{l-1}y + \ldots + a_{l0}y^l).$$

Thus such a polynomial has at most $1+2+3+\ldots+(l+1) = \frac{1}{2}(l+1)(l+2)$ arbitrary constants a_{ij}. When this is substituted into Laplace's equation, the double differentiation yields an equation that asserts that a certain homogeneous polynomial of degree $(l-2)$ must vanish identically. In other words, each of the $\frac{1}{2}l(l-1)$ coefficients in that second polynomial must separately vanish. This establishes $\frac{1}{2}l(l-1)$ relations between the original $\frac{1}{2}(l+1)(l+2)$ coefficients, and thus finally leaves only $\frac{1}{2}(l+1)(l+2) - \frac{1}{2}l(l-1) = 2l+1$ of the coefficients independent.

the corresponding radial dependence of the wave functions will be given by appropriate solutions to (5.5), with $k = l(l+1)$.

Suppose you have found several H_l for some value of l. Then any linear combination of them is also an H_l, because (5.10) is a linear equation. And since any such combination has the same value of l, and therefore of k in (5.5) for R, the radial dependence of the wave functions for all these H_l is the same for a given value of l. Furthermore the permitted values of E, the energy levels, are the same for a given value of l, since E appears only in (5.5), not in (5.6).

Notice now several implications of these facts. In the first place, you have found 'all' the H_l when (1) you have found a set that is linearly independent, and (2) you cannot find another H_l that is linearly independent of those already found. In the second place, the wave functions for a given value of l are degenerate:† several, with different H_l, have the same energy levels. If M linearly independent H_l can be found for a given l, then the wave functions for the value l have M-fold degeneracy.

Now when you have the permitted solutions of (5.5) for a given value of l, you can number the corresponding energy levels by a number n, in order of decreasing negative E: $E_{n,l}$. Since each energy level is M-fold degenerate, you can index the wave functions by using a third number m: $\psi_{n,l,m}$.

Suitable sets of H_l for small values of l are easily found by trial. You can ignore the methods for systematically exhibiting such sets, since higher values of l will seldom concern you. When $l = 0$, the polynomial that satisfies (5.10) is a constant—any constant. When $l = 1$, the coordinates x, y, and z satisfy (5.10), and hence any linear function of x, y, and z also satisfies it.

† For a discussion of degeneracy, see *Stationary states*, Chapter 6.

It is convenient to pick x, y, and z themselves as the complete set. When $l = 2$, a suitable set is xy, yz, zx, x^2-y^2, $2z^2-x^2-y^2$. Notice that the solution z^2-y^2 is not linearly independent of these: it is half the sum of the last two. For $l = 3$, the polynomials xyz, $x(y^2-z^2)$, $y(z^2-x^2)$, $z(x^2-y^2)$, $x(x^2-3y^2)$, $y(y^2-3z^2)$, $z(z^2-3x^2)$, form a suitable set.

Notice that $2l+1$ functions are shown for each value of l. In fact, as Discussion 5.2 demonstrates, there must always be just $2l+1$ linearly independent H_l that satisfy Laplace's equation (5.10).

It is sometimes useful to place on a set of solutions for a given value of l the additional requirement that its members be *orthogonal* to one another. Out of a set of linearly independent wave functions that are not all orthogonal, a set of an equal number that *are* orthogonal can always be constructed.† For example, the set just listed for $l = 2$ is completely orthogonalized, but the set for $l = 3$ is not.

One of the advantages of solving the wave equation for this problem in spherical polar coordinates, instead of in Cartesian coordinates as this chapter has done, is that the polar method‡ automatically provides wave functions that are orthogonal. Another advantage is that it offers a systematic procedure for exhibiting a complete set of wave functions for all values of l. The fact that the wave functions exhibited by the polar method give special importance to one direction is an advantage when such a direction has physical significance and a disadvantage when it does not. In many problems of atomic spectroscopy, the direction of an applied magnetic field gives significance to an axis. In some problems of chemistry, when an atom is bonded to one other atom in a molecule, for example, a natural axis appears. On the other hand, in the simplest situations in solids, such an axis is illusory. But of course the wave functions provided by either method are essentially equivalent: any set exhibited by one method can be constructed by taking linear combinations of the members of any set exhibited by the other method.

In summary, wave functions for an electron in a central field can be chosen to have the form $R_l H_l/r^l$, where H_l/r^l is a function of direction alone and R_l is a function of distance from the centre. When $l = 0$, the wave function is independent of direction. When $l = 1$, the three choices of H_l proposed above give the directional dependences x/r, y/r, and z/r, or, in other words, the cosine of the angle between the direction and the axes of x, y, and z respectively. Fig. 5.3 shows what the squares of these directional dependences look like: the angular dependence of the relative probability of finding the electron, or the 'density of the electron cloud' as a function of angle. Do not make the mistake of thinking that these diagrams picture the electron cloud

† The idea of orthogonality, and the method for constructing an orthogonal set, are discussed in *Stationary states*, Chapter 6.

‡ See for example L. Pauling and E. B. Wilson, *Introduction to quantum mechanics*, McGraw-Hill, New York, 1935.

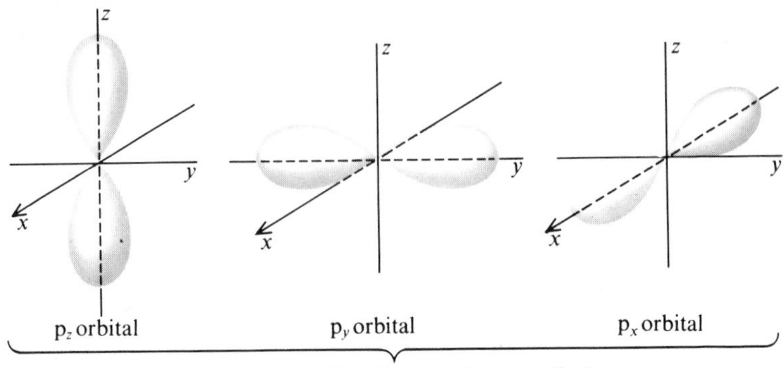

FIG. 5.3. The angular factor for the probability cloud of a particle moving in a central force field, in a choice of the three wave functions for $l = 1$ in which the x, y, and z axes are treated alike.

itself. They do not take account of the radial dependence; they show the angular factor by which that radial dependence must be multiplied. They show, for instance, that the three choices of H_l for $l = 1$ give a maximum probability along an axis and a probability of zero perpendicular to that axis, and that the probability is rotationally symmetrical around that axis.

PROBLEMS

5.1 Show that the sum of the three probability density functions for $l = 1$ is spherically symmetrical. The fact that the sum of the $2l+1$ probability density functions is spherically symmetrical for any value of l is *Unsöld's Theorem.*

5.2 Since the problem that this chapter has treated is spherically symmetrical, all the procedures (up to the point where special H_l are chosen) are expressible in vector form. Write (5.1), (5.2), (5.4), (5.6), (5.7), (5.8), (5.9), and (5.10) in vector notation.

5.3 Many one-particle problems are not spherically symmetric but are *axially symmetric*: the potential for the particle is a function of its *distance from* the axis and of its *position along* the axis. The problem of an electron attracted to two fixed nuclei in a diatomic molecule is one important example. Letting z be the axis and $r = \sqrt{(x^2+y^2)}$ the distance from the axis, you can seek wave functions of the form $\psi = R(z, r)L(x/r, y/r)$ by analogy with the methods of this chapter. Show that in such a wave function, the parts must satisfy the equations

$$(1)\quad \frac{\partial^2 R}{\partial r^2}+\frac{\partial^2 R}{\partial z^2}+\frac{1}{r}\frac{\partial R}{\partial r}+\left[E-V(z, r)-\frac{l^2}{r^2}\right]R = 0,$$

$$(2)\quad \frac{\partial^2 H_l}{\partial x^2}+\frac{\partial^2 H_l}{\partial y^2} = 0,$$

where $L = H_l/r^l$ and H_l is a homogeneous polynomial of order l in x and y.

5.4 Noticing that $(x+iy)^m+(x-iy)^m = H_m$ fulfills the requirements of Problem 5.3, show that there are wave functions for the axially symmetric problem whose dependence on the angle ϕ about the axis is $L \sim \cos m\phi$ where m is any integer up to and including l.

5.5 Give in words a simple reason why the results of Problems 5.3 and 5.4 must be valid also for the spherically symmetric problem.

6. Angular Momenta of a Particle in a Central Field

TURN now to consider the angular momenta of a particle, such as an electron, in the states whose wave functions $\psi = RL$ were examined in the last chapter. You will see that you can pick functions for which the energy, the total angular momentum, and the angular momentum about one chosen axis, are all simultaneously specified. But when these three quantities have been specified, the limitations imposed by the uncertainty principle permit you only to calculate an expectation value of the angular momentum about any other axis. You may be surprised to find that the angular factors L contain all the information obtainable about angular momentum: that property is independent of the radial factors R.

Since the procedure of the last chapter introduced no unique axis, examine first the *total angular momentum*. Angular momentum is a vector quantity, and *total angular momentum* means the absolute value of that vector, its length, a quantity that is independent of axes. You will discover that each of the angular factors L derived in the last chapter has a precisely specified value of total angular momentum, which depends only on l. The total angular momentum is therefore quantized, like the energy, for l can be only an integer.

Then you can go on to choose an axis of special interest and seek a wave function that exhibits a precisely specified value of the angular momentum about that axis. The last chapter showed that the wave functions for a particular value of l exhibit $(2l+1)$-fold degeneracy. Thus you can seek a linear combination of the functions of given l that provides a specified value of one component of angular momentum, and you will find that this component is also quantized. This chapter is devoted to showing how to carry out the suggested programme.

The wave-mechanical operators† for the three components of angular momentum, expressed in Cartesian coordinates, are

$$M_x = \frac{h}{2\pi i}\left(y\frac{\partial}{\partial z} - z\frac{\partial}{\partial y}\right),$$

$$M_y = \frac{h}{2\pi i}\left(z\frac{\partial}{\partial x} - x\frac{\partial}{\partial z}\right),$$

$$M_z = \frac{h}{2\pi i}\left(x\frac{\partial}{\partial y} - y\frac{\partial}{\partial x}\right).$$

† These operators are described in *Stationary states*, Chapter 8.

Since the angular momentum is a vector, the square of the total angular momentum is the sum of the squares of its three components:

$$M^2 = M_x^2 + M_y^2 + M_z^2.$$

To 'square' an operator means to operate with it twice in succession. The appendix to this chapter carries out the formalities of operating with M^2 on the wave function $\psi = RL$.

The result of this investigation is that the square of the total angular momentum of an electron, when its wave function has the quantum number l, has the exact value $l(l+1)(h/2\pi)^2$, no matter what sort of well the electron is in, so long as that well has spherical symmetry. An electron in a state with $l = 1$ for example, has the total angular momentum $\sqrt{(2)}h/2\pi$; an electron for which $l = 2$ has $\sqrt{(6)}h/2\pi$. You cannot know what the direction of that angular momentum is, but you can specify the value of the component along some one direction by choosing a particular angular factor L out of the sets that can be constructed from the H_l.

In undertaking that task, it is clear at once that the maximum possible value about any one axis is $\sqrt{\{l(l+1)\}}h/2\pi$, since the angular momentum about one axis could not exceed the total angular momentum. Furthermore even that maximum could not be specified exactly. If it were, the specification would assert: 'The angular momentum is all around that axis, and therefore the angular momentum about all directions perpendicular to that axis is precisely zero'. But the uncertainty principle asserts that it must be impossible to specify all the components of angular momentum at once.

You will find that the largest absolute value of angular momentum that can be specified about any one axis is $lh/2\pi$. Moreover it turns out that the only precisely determined values that the angular momentum can take about one axis run from $-lh/2\pi$ to $+lh/2\pi$, in integral multiples of $h/2\pi$. In other words, the values of angular momentum along any axis are quantized, whenever there is a way to fix them accurately.

To see how this quantization comes about, pick the favoured axis as the z axis in a system of Cartesian coordinates. The quantum-mechanical operator for the angular momentum about one axis is $M_z = h/2\pi i(x\,\partial/\partial y - y\,\partial/\partial x)$. To find a wave function in which that angular momentum has a precise value requires that the result of applying the operator to the wave function shall be equivalent to multiplying the wave function by a constant. The wave function is then said to be an *eigenfunction of the operator*, and the constant to be an *eigenvalue of the operator*. The requirement can be written

$$M_z\psi = \frac{h}{2\pi i}\left(x\frac{\partial\psi}{\partial y} - y\frac{\partial\psi}{\partial x}\right) = m\psi, \qquad (6.1)$$

where m is a constant.

Substituting the wave function $\psi = RL$ in (6.1) yields

$$\frac{h}{2\pi i}L\left(x\frac{\partial R}{\partial y}-y\frac{\partial R}{\partial x}\right)+\frac{h}{2\pi i}R\left(x\frac{\partial L}{\partial y}-y\frac{\partial L}{\partial x}\right) = mRL. \tag{6.2}$$

Since $\partial R/\partial y = y/r\, \mathrm{d}R/\mathrm{d}r$, etc., the first term on the left vanishes, and the equation becomes

$$\frac{h}{2\pi i}\left(x\frac{\partial L}{\partial y}-y\frac{\partial L}{\partial x}\right) = mL. \tag{6.3}$$

Again, therefore, the results of the inquiry will not depend on the radial form of the wave function.

The substitution of $L = H_l/r^l$ yields

$$\frac{h}{2\pi i}\left(x\frac{\partial L}{\partial y}-y\frac{\partial L}{\partial x}\right) = \frac{h}{2\pi i}\left[\frac{x}{r^l}\frac{\partial H_l}{\partial y}-\frac{y}{r^l}\frac{\partial H_l}{\partial x}-\frac{lH_l}{r^{l+2}}(xy-yx)\right] = \frac{mH_l}{r^l}, \tag{6.4}$$

or

$$\frac{h}{2\pi i}\left(x\frac{\partial H_l}{\partial y}-y\frac{\partial H_l}{\partial x}\right) = mH_l. \tag{6.5}$$

In other words you now require not only that the H_l be solutions to Laplace's equation but that they be solutions to (6.5).

Content yourself with examining how this works for the first few values of l. Of course, when $l = 0$ and H_l is a constant, $m = 0$ is the only solution: such functions have no angular momentum about any axis. This may sound at first as if it contradicted the earlier discussion of the uncertainty principle. But in this case angular momentum simply does not come into question, just as it does not in the one-dimensional problems discussed earlier, for the *total* angular momentum is zero.

When $l = 1$, any linear combination of x, y, and z gives an admissible function. Substituting $H_l = ax+by+cz$ in (6.5) yields

$$\frac{h}{2\pi i}(bx-ay) = m(ax+by+cz). \tag{6.6}$$

In order to satisfy (6.6) identically, the coefficients of x, y, z must separately vanish:

$$\frac{h}{2\pi i}b = ma, \qquad \frac{-h}{2\pi i}a = mb, \qquad 0 = mc. \tag{6.7}$$

The first two members of (6.7) yield $m^2 = (h/2\pi)^2$, or $m = \pm h/2\pi$. If $m = +h/2\pi$, then $b = \mathrm{i}a$; if $m = -h/2\pi$, then $b = -\mathrm{i}a$. In both cases $c = 0$. If $m = 0$, then $a = b = 0$ and c is arbitrary. Notice that in all three cases one arbitrary constant remains, which can be used to normalize the wave functions. The three functions quantized along the z axis are then

$$AR\left(\frac{x}{r}+\mathrm{i}\frac{y}{r}\right), \qquad AR\left(\frac{x}{r}-\mathrm{i}\frac{y}{r}\right), \qquad AR\frac{z}{r}, \tag{6.8}$$

where A is the arbitrary constant. In terms of spherical coordinates with their poles on the z axis, $x = r \sin\theta \cos\phi$, $y = r \sin\theta \sin\phi$, $z = r\cos\theta$, and (6.8) transforms to

$$AR\mathrm{e}^{+\mathrm{i}\phi}\sin\theta, \qquad AR\mathrm{e}^{-\mathrm{i}\phi}\sin\theta, \qquad AR\cos\theta. \tag{6.9}$$

To show the appearance of these angular functions Fig. 6.1 is given for

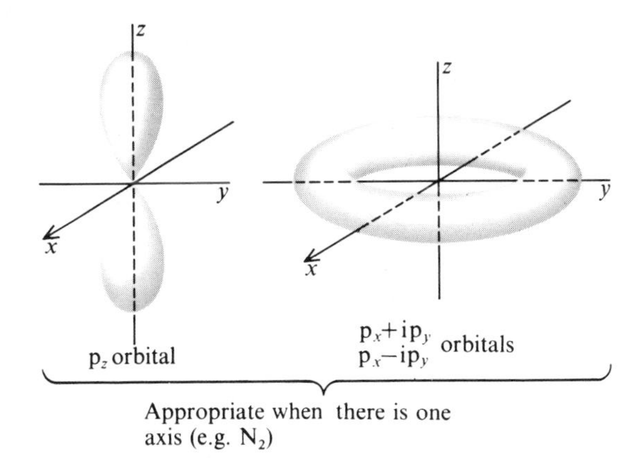

FIG. 6.1. The angular factor for the probability cloud of a particle moving in a central force field, in a choice of the three wave functions for $l = 1$ in which the angular momentum about the z axis is given precisely specified values.

comparison with Fig. 5.3. Remember that probability distributions are measured by the product of a wave function and its complex conjugate. Hence the angular factors for the first two distributions are the same: $\sin^2\theta$. Since the two functions refer to angular momenta with opposite signs, the sign in the exponential factor signifies which direction the electron takes in moving around the z axis. Notice that the earlier choice of functions with $l = 1$, namely Rx/r, Ry/r, Rz/r, was a choice in which the angular momentum is zero about the x, y, and z axes respectively. Discussion 6.1 provides the outlines of a similar discussion for states with $l = 2$.

Since the angular momentum about the axis is always an integral multiple of $h/2\pi$, that multiplier is a natural quantum number with which to index the functions for a given value of l. The number m takes $2l+1$ integral values: the positive and negative integers from $-l$ to $+l$ including zero. The quantum

Discussion 6.1

MAGNETIC QUANTUM NUMBERS WHEN $l = 2$

To extend to the wave functions with $l = 2$ the reasoning exemplified in the text for the functions with $l = 1$, write the most general H_2 in the form

$$H_2 = a_0 z^2 + z(a_1 x + b_1 y) + a_2 x^2 + 2b_2 xy + c_2 y^2. \tag{D6.1}$$

Substituting this in Laplace's equation gives the requirement

$$a_0 + a_2 + c_2 = 0. \tag{D6.2}$$

Substituting it also in (6.5) gives the requirements

$$0 = ma_0, \quad \frac{h}{2\pi \mathrm{i}} b_1 = ma_1, \quad 2\frac{h}{2\pi \mathrm{i}} b_2 = ma_2,$$

$$\frac{-h}{2\pi \mathrm{i}} a_1 = mb_1, \quad -2\frac{h}{2\pi \mathrm{i}} b_2 = mc_2, \tag{D6.3}$$

$$2\frac{h}{2\pi \mathrm{i}}(c_2 - a_2) = 2mb_2.$$

From (D6.2) and (D6.3) the conditions on H_2 for quantization along the z axis become:

$$m = 0: \quad c_2 = a_2 = \frac{-a_0}{2}, \quad a_1 = b_1 = b_2 = 0.$$

$$m = \frac{+h}{2\pi}: \quad b_1 = +\mathrm{i}a_1, \quad a_0 = a_2 = b_2 = c_2 = 0.$$

$$m = \frac{-h}{2\pi}: \quad b_1 = -\mathrm{i}a_1, \quad a_0 = a_2 = b_2 = c_2 = 0.$$

$$m = +2\frac{h}{2\pi}: \quad a_2 = -\mathrm{i}b_2, \quad c_2 = +\mathrm{i}b_2, \quad a_0 = a_1 = b_1 = 0.$$

$$m = -2\frac{h}{2\pi}: \quad a_2 = +\mathrm{i}b_2, \quad c_2 = -\mathrm{i}b_2, \quad a_0 = a_1 = b_1 = 0.$$

Thus by using these coefficients in the general form (D6.1), five wave functions are constructed, for each of which the component of angular momentum along the z axis has a precisely specified value.

number l is called the *azimuthal quantum number* of the state in question; m is called the *magnetic quantum number*. And m acquires significance only when the physical situation gives significance to a unique axis.

PROBLEMS

6.1 When the electron in a hydrogen atom is in a state of maximum m about the z axis, for a given l, what is the value of the component of its orbital angular momentum in the plane perpendicular to z?

6.2 Show that the first two wave functions of (6.9) are orthogonal† over the accessible values of θ and ϕ.

6.3 Find the values of the constant A that normalize† the wave functions of (6.9), assuming that R is already normalized.

6.4 (a) Show that for the *axially symmetric* problem (Problem 5.3) eqn (6.3) still applies.
(b) Use wave functions of the type of Problem 5.4 to show that for any value of l there is a wave function in which the angular momentum M_z has the precise value $\pm m(h/2\pi)$, where m is any integer up to and including l.

6.5 By transforming the coordinate system for the axially symmetric problem to cylindrical coordinates r, ϕ, z (where $x = r\cos\phi, y = r\sin\phi, z = z$), show that the operator $M_z = h/2\pi\mathrm{i}[x(\partial/\partial y) - y(\partial/\partial x)]$ transforms into the operator $M_z = (h/2\pi\mathrm{i})(\partial/\partial\phi)$.

APPENDIX

TOTAL ANGULAR MOMENTUM

The wave-mechanical operators for the components of angular momentum about the x, y, and z axes are

$$M_x = \frac{h}{2\pi\mathrm{i}}\left(y\frac{\partial}{\partial z} - z\frac{\partial}{\partial y}\right),$$

$$M_y = \frac{h}{2\pi\mathrm{i}}\left(z\frac{\partial}{\partial x} - x\frac{\partial}{\partial z}\right),$$

$$M_z = \frac{h}{2\pi\mathrm{i}}\left(x\frac{\partial}{\partial y} - y\frac{\partial}{\partial x}\right).$$

Since $M^2 = M_x^2 + M_y^2 + M_z^2$, the squares of the operators for the three components must be applied to the wave functions, and the results must be added together, in order to examine the square of the total angular momentum. A typical operation is

$$M_z^2\psi = -\left(\frac{h}{2\pi}\right)^2\left(x\frac{\partial}{\partial y} - y\frac{\partial}{\partial x}\right)\left(x\frac{\partial}{\partial y} - y\frac{\partial}{\partial x}\right)\psi$$

$$= \left(\frac{h}{2\pi}\right)^2\left[2xy\frac{\partial^2\psi}{\partial x\,\partial y} - x^2\frac{\partial^2\psi}{\partial y^2} - y^2\frac{\partial^2\psi}{\partial x^2} + x\frac{\partial\psi}{\partial x} + y\frac{\partial\psi}{\partial y}\right].$$

The sum of the three terms of this sort yields

$$\left(\frac{2\pi}{h}\right)^2 M^2\psi = \left(x^2\frac{\partial^2\psi}{\partial x^2} + y^2\frac{\partial^2\psi}{\partial y^2} + z^2\frac{\partial^2\psi}{\partial z^2} + 2yz\frac{\partial^2\psi}{\partial y\,\partial z} + 2zx\frac{\partial^2\psi}{\partial z\,\partial x} + 2xy\frac{\partial^2\psi}{\partial x\,\partial y}\right) +$$

$$+2\left(x\frac{\partial\psi}{\partial x} + y\frac{\partial\psi}{\partial y} + z\frac{\partial\psi}{\partial z}\right) - r^2\left(\frac{\partial^2\psi}{\partial x^2} + \frac{\partial^2\psi}{\partial y^2} + \frac{\partial^2\psi}{\partial z^2}\right). \quad \text{(A6.1)}$$

To make further work easier, notice that the square of the operator $x\,\partial/\partial x + y\,\partial/\partial y + z\,\partial/\partial z$ is

$$\left(x\frac{\partial}{\partial x} + y\frac{\partial}{\partial y} + z\frac{\partial}{\partial z}\right)^2 = x\frac{\partial}{\partial x} + y\frac{\partial}{\partial y} + z\frac{\partial}{\partial z} + x^2\frac{\partial^2}{\partial x^2} + y^2\frac{\partial^2}{\partial y^2} + z^2\frac{\partial^2}{\partial z^2} +$$

$$+2yz\frac{\partial^2}{\partial y\,\partial z} + 2zx\frac{\partial^2}{\partial z\,\partial x} + 2xy\frac{\partial^2}{\partial x\,\partial y}. \quad \text{(A6.2)}$$

† See *Stationary states*, Chapters 2 and 6.

This permits rewriting the operator for M^2 in (A6.1) in the form

$$\left(\frac{2\pi}{h}\right)^2 M^2 = \left(x\frac{\partial}{\partial x}+y\frac{\partial.}{\partial y}+z\frac{\partial}{\partial z}\right)^2+\left(x\frac{\partial}{\partial x}+y\frac{\partial}{\partial y}+z\frac{\partial}{\partial z}\right)-r^2\left(\frac{\partial^2}{\partial x^2}+\frac{\partial^2}{\partial y^2}+\frac{\partial^2}{\partial z^2}\right)$$

$$= \left(x\frac{\partial}{\partial x}+y\frac{\partial}{\partial y}+z\frac{\partial}{\partial z}+1\right)\times\left(x\frac{\partial}{\partial x}+y\frac{\partial}{\partial y}+z\frac{\partial}{\partial z}\right)-r^2\left(\frac{\partial^2}{\partial x^2}+\frac{\partial^2}{\partial y^2}+\frac{\partial^2}{\partial z^2}\right). \tag{A6.3}$$

Now operate on the wave function RL with this operator. Since RL is a solution of Schrödinger's equation (5.1), the second term of the operation produces $r^2(E-V)RL$. Operating with the second factor of the first term yields

$$\left(x\frac{\partial}{\partial x}+y\frac{\partial}{\partial y}+z\frac{\partial}{\partial z}\right)RL = R\left(x\frac{\partial L}{\partial x}+y\frac{\partial L}{\partial y}+z\frac{\partial L}{\partial z}\right)+L\left(x\frac{\partial R}{\partial x}+y\frac{\partial R}{\partial y}+z\frac{\partial R}{\partial z}\right). \tag{A6.4}$$

The first term on the right of (A6.4) vanishes, by Euler's theorem, because L is homogeneous of order zero. In consequence of the relations

$$\partial R/\partial x = (\mathrm{d}R/\mathrm{d}r)(\partial r/\partial x) = (x/r)(\mathrm{d}R/\mathrm{d}r), \text{ etc.,}$$

the second term on the right of (A6.4) equals $2r(\mathrm{d}R/\mathrm{d}r)$. Thus (A6.3) can be written

$$\left(\frac{2\pi}{h}\right)^2 M^2RL = r^2(E-V)RL+\left(x\frac{\partial}{\partial x}+y\frac{\partial}{\partial y}+z\frac{\partial}{\partial z}+1\right)Lr\frac{\mathrm{d}R}{\mathrm{d}r}. \tag{A6.5}$$

Perform now the remaining operation on the right of (A6.5):

$$\left(x\frac{\partial}{\partial x}+y\frac{\partial}{\partial y}+z\frac{\partial}{\partial z}+1\right)Lr\frac{\mathrm{d}R}{\mathrm{d}r} = \left(x\frac{\partial L}{\partial x}+y\frac{\partial L}{\partial y}+z\frac{\partial L}{\partial z}\right)r\frac{\mathrm{d}R}{\mathrm{d}r}+$$
$$+\left(x\frac{\partial r}{\partial x}+y\frac{\partial r}{\partial y}+z\frac{\partial r}{\partial z}\right)L\frac{\mathrm{d}R}{\mathrm{d}r}+$$
$$+\left(x\frac{\partial}{\partial x}\frac{\mathrm{d}R}{\mathrm{d}r}+y\frac{\partial}{\partial y}\frac{\mathrm{d}R}{\mathrm{d}r}+z\frac{\partial}{\partial z}\frac{\mathrm{d}R}{\mathrm{d}r}\right)Lr+$$
$$+Lr\frac{\mathrm{d}R}{\mathrm{d}r}. \tag{A6.6}$$

Again the first term on the right of (A6.6) vanishes, by Euler's theorem. The second term equals $Lr(\mathrm{d}R/\mathrm{d}r)$, and the third term equals $Lr^2(\mathrm{d}^2R/\mathrm{d}r^2)$. Eqn (A6.5) thus becomes

$$\left(\frac{2\pi}{h}\right)^2 M^2RL = r^2(E-V)RL+2Lr\frac{\mathrm{d}R}{\mathrm{d}r}+Lr^2\frac{\mathrm{d}^2R}{\mathrm{d}r^2}. \tag{A6.7}$$

But since R satisfies (5.5), the right side of (A6.7) equals kRL, and that equation can be written

$$M^2RL = \left(\frac{h}{2\pi}\right)^2 kRL. \tag{A6.8}$$

Finally, since k must be equal to $l(l+1)$ as the last chapter showed, the square of the total angular momentum can be given the exact value $l(l+1)(h/2\pi)^2$.

7. Hydrogenlike Atomic States

WHEN there is only one electron in an atom, it finds itself in a spherically symmetrical well, formed by the attraction of the oppositely charged nucleus. For such an electron, Chapter 5 showed how the radial dependence of the wave function could be separated from its angular dependence. Using the form $\psi = RL$, you can find angular factors L that are independent of the radial dependence R of the wave function. In other words, the angular factors do not depend on the precise sort of well the electron is in, so long as it is spherically symmetrical—so long as the instantaneous potential energy V is a function of r only.

But the radial factors depend on the angular factors, because they depend on the value of the total angular momentum. The differential equation obeyed by the radial factors (5.5) is

$$\frac{\mathrm{d}^2 R}{\mathrm{d}r^2}+\frac{2}{r}\frac{\mathrm{d}R}{\mathrm{d}r}+\left[E-\frac{l(l+1)}{r^2}-V(r)\right]R = 0, \tag{7.1}$$

expressed in atomic units. The quantum number l can take all positive integral values, including zero, and the total angular momentum has the value $\sqrt{\{l(l+1)\}}h/2\pi$, as the last chapter showed.

It is helpful to make the transformation $R = \rho/r$ in (7.1), so that the dependent variable in the equation becomes rR. This eliminates the second term in (7.1) without changing anything else, so that it reads

$$\frac{\mathrm{d}^2(rR)}{\mathrm{d}r^2}+\left[E-\frac{l(l+1)}{r^2}-V(r)\right](rR) = 0. \tag{7.2}$$

This looks like a one-dimensional wave equation for an electron with the potential energy function $V(r)+l(l+1)/r^2$ instead of just $V(r)$. Since the true potential energy $V(r)$ is negative, the angular momentum acts to reduce the depth of the potential well. The total attractive force from the nucleus can be pictured as partly counterbalanced by the centrifugal force on the electron, to a greater extent as the angular momentum increases.

The task at hand is to find the appropriate solutions to (7.2) when the potential energy is that of an electron in the presence of a nucleus of charge Ze: $V(r) = -Ze^2/4\pi\epsilon_0 r$. Of course the only electrically neutral atom for which V takes this form is the hydrogen atom, where $Z = 1$. But the helium ion, the helium atom that has lost one electron, realizes the case $Z = 2$, and the analysis is just as easily carried out for any value of Z. The resulting model is called *the hydrogenlike atom.*

Since (7.2) is expressed in atomic units (Discussion 4.1), the transformation $r \to (h^2\epsilon_0/\pi me^2)r$ must be applied to $V(r)$, yielding $(-2Z/r)(me^4/8h\epsilon_0^2)$, so that $V(r) = -2Z/r$ in atomic units. Then (7.2) reads

$$\frac{d^2(rR)}{dr^2}+\left[E-\frac{l(l+1)}{r^2}+\frac{2Z}{r}\right](rR) = 0. \tag{7.3}$$

The appendix to this chapter carries out the investigation necessary to develop the appropriate solutions to this equation. The analysis shows that the energy of the electron in a bound state is quantized: it can take on only the values $-E = Z^2/n^2$, where n is a whole number equal to or greater than $l+1$. The lowest admissible energy, corresponding to the *ground state* of the atom, is therefore $E = -Z^2$, when $l = 0$ and $n = 1$.

If the electron has any of the other possible energies, the atom is said to be in an *excited state*. The electron will tend to drop back to a lower level, radiating an energy ΔE equal to the difference between the two levels. The frequency ν, and thus the wavelength of the radiation, will be given by $\Delta E = h\nu$. By observing the resulting spectra, the possible energy levels are determined experimentally. Not all interlevel transitions can actually occur; the possible transitions are limited by *selection rules*, which are again deducible by quantum mechanics but which this chapter will not discuss.

Before the quantum mechanical explanation of the spectral lines was available, spectroscopists had classified them into series, and had given these series the names *sharp, principal, diffuse,* and *fundamental.* The initial letters of these words have been retained to describe the states and their associated energy levels in which $l = 0$ (s), $l = 1$ (p), $l = 2$ (d), and $l = 3$ (f). For larger values of l, the letters g, h, etc., are used in succession. The *upper* level of the two between which a radiative transition occurs to give a line of the *sharp* series for example, is an 's level', belonging to an 's state', with $l = 0$.

Thus there is a series of s states in which $n = 1, 2, 3$, etc., and $-E = Z^2$, $(\frac{1}{4})Z^2$, $(\frac{1}{9})Z^2$, etc.; a series of p states in which $n = 2, 3$, etc.; a series of d states in which n starts with 3; and so on. The quantum number n is called the *principal* quantum number, and l is called the *azimuthal* quantum number. A state is described as a '1s state', a '3p state', etc., by naming its principal quantum number and then the appropriate spectroscopic letter for the value of l.

Notice further that for a given value of the principal quantum number n, and thus of the energy, there are one s state (with $l = 0$), three p states (with $l = 1$), five d states (with $l = 2$), and so on, up to $2n-1$ states with $l = n-1$. In short there are n^2 states with an energy corresponding to the principal quantum number n; the states of the hydrogenlike atom have n^2-fold degeneracy. It is useful to keep in mind a table (Fig. 7.1) of the number of states corresponding to the successive values of n with different values of l. As n increases and the negative energy decreases, there are more and

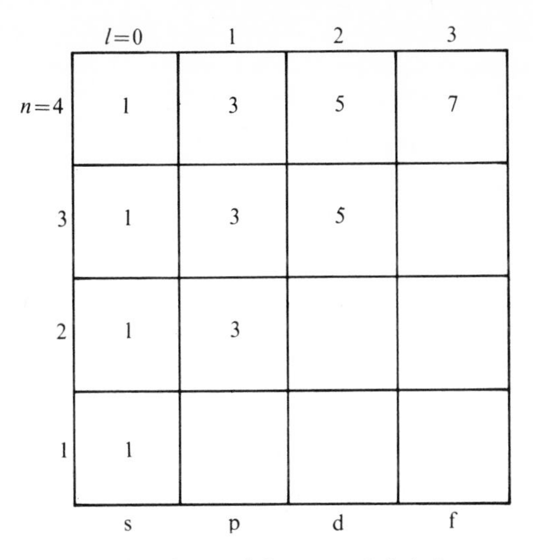

FIG. 7.1. The number of states of given energy (principal quantum number n) with various values of total angular momentum (azimuthal quantum number l). Each state can accomodate two electrons.

more states with the same energy (numbered by n) and with different values of angular momentum (numbered by l and m).

The discussion in Chapter 5 has already shown that there is a $(2l+1)$-fold degeneracy of states with the same value of l. That degeneracy arises from the fact that the potential well has spherical symmetry, and does not depend on other aspects of the shape of the well. The additional degeneracy exhibited by the argument of this chapter *does* depend on the shape of the well, on the fact that $V(r)$ has the form $-(Ze^2/4\pi\epsilon_0 r)$. Atoms that are hydrogenlike, such as hydrogen itself and the helium ion, actually exhibit this additional degeneracy. But other atoms, with more than one electron, do not, for reasons that the next chapter will discuss. Fig. 7.2 shows the lowest energy levels for hydrogen, and the allowed transitions and spectroscopic states.

Examine now the form of the radial part of the wave functions for these states. As the appendix shows, that form is given by (A7.6),

$$R = e^{-\alpha r} r^l (a_0 + a_1 r + \ldots + a_{j-1} r^{j-1}), \tag{A7.6}$$

where $\alpha^2 = -E$. The coefficients obey the recursion relation

$$a_j = \frac{-2}{j} \cdot \frac{Z - \alpha(l+j)}{2l+j+1} a_{j-1} \tag{A7.5}$$

and the polynomial is terminated by taking $Z = \alpha(l+j)$. Notice that in all the preceding members of the recursion relation, Z will be greater than the

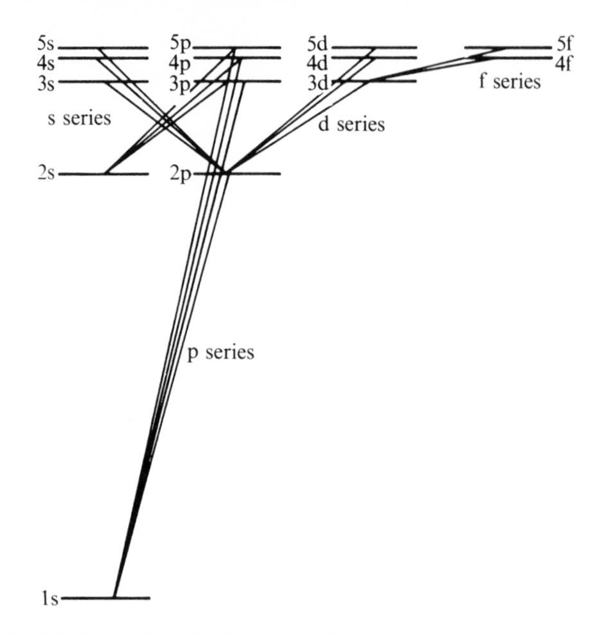

FIG. 7.2. Energy levels in the hydrogenlike atom, and radiative transitions between those levels that produce the main observed series of lines in the emission spectra of the atoms.

other term in the numerator, and hence the signs of the coefficients in the polynomial in (A7.6) will be alternately plus and minus. Descartes' rule of signs asserts that such a polynomial with j terms will have $j-1$ real positive roots, or fewer by an even number, and that if they are fewer, the other roots will be complex. It turns out that the roots of these polynomials are all real, and hence the radial part of the wave function oscillates with increasing r, passing through zero $j-1$ times, in other words $n-l-1$ times, since $n = l+j$.

The radial dependence of the wave functions for the lowest three s states is shown in Fig. 7.3. They have an increasing number of zeros as n increases, in agreement with the last paragraph.† Despite the increasing number of oscillations, however, the mean kinetic energy in *all* the wave functions, not only the s functions, decreases as n increases. The increased distance over which the wave function extends more than compensates for its increased number of zeros. The total energy still increases under these circumstances because the mean potential energy increases even more rapidly than the kinetic energy decreases. The electron spends more of its time farther from the nucleus.

† This fact is also in agreement with the discussion of orthogonality in *Stationary states*, Chapter 6.

There are two interesting ways of looking at the probability distributions given by these wave functions. One way is to look directly at their squares, which will measure the relative probabilities of finding the electron in little volume elements, all of the same size, at various distances from the nucleus. The other way is to look at the square of the wave function multiplied by $4\pi r^2$. That will measure the relative probability of finding the electron somewhere in a thin spherical shell of radius r, or, in other words, the probability of finding it at the distance r from the nucleus.

It is clear in Fig. 7.3 that the s functions give a maximum at $r = 0$ for the first kind of probability: the electron is more likely to be at the nucleus than anywhere else. Fig. 7.4 shows curves of the second kind of probability. As n increases, the most probable distance of the electron from the nucleus moves outward. The two figures together suggest, thinking roughly, the electron distribution in a 1s state as a spherical ball, in a 2s state as a spherical ball, surrounded by an outer shell, and in a 3s state as a ball with two outer shells.

When l is not zero, the wave functions are not spherically symmetrical, and it is more difficult to visualize them because both their radial and their angular dependence must be considered. In the case of the p functions, the square of that angular dependence, in other words, the angular dependence of the corresponding probability functions, has already been shown in Fig. 5.3 and Fig. 6.1. The radial dependence of the 2p and 3p functions is plotted in Fig. 7.5. The squares of these functions show that, unlike the s functions, the p functions do not afford a maximum probability of finding the electron at the nucleus. In fact for all states in which l is not zero, the probability of finding the electron at the nucleus vanishes, because the wave functions have the factor r^l (A7.6).

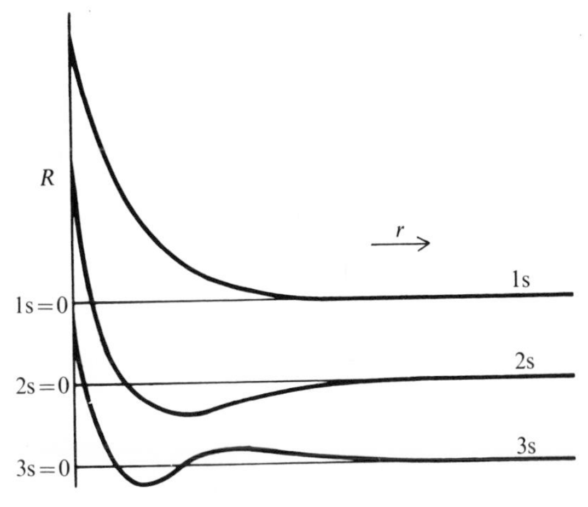

FIG. 7.3. Radial dependence of the first three s functions ($l = 0$) in the hydrogenlike atom.

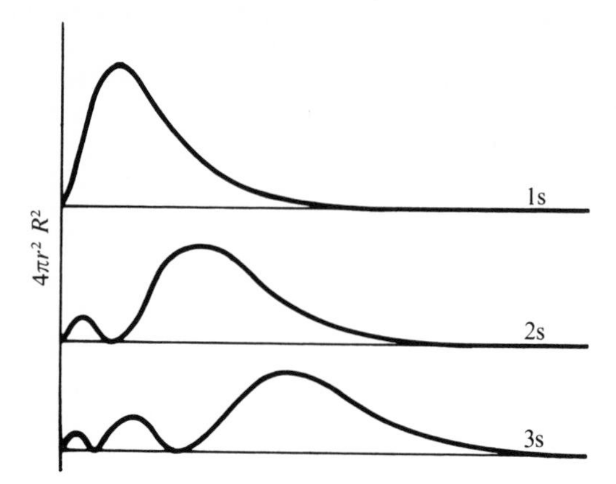

FIG. 7.4. Probability of finding the electron in a thin spherical shell at the distance r from the nucleus, in the first three s functions of the hydrogenlike atom.

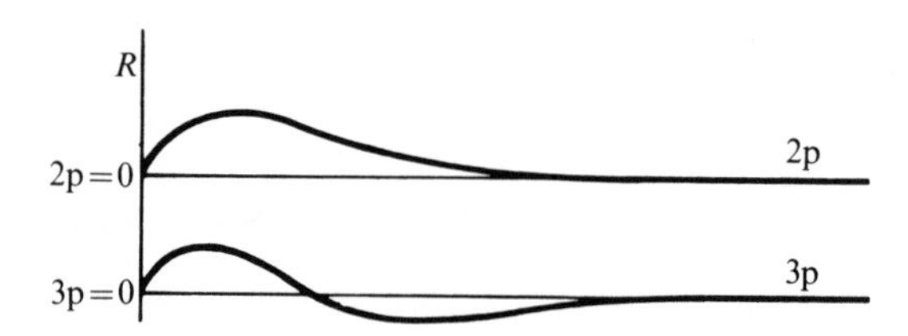

FIG. 7.5. Radial dependence of the first two p functions ($l = 1$, total angular momentum $\sqrt{(2)}h/2\pi$) of the hydrogenlike atom.

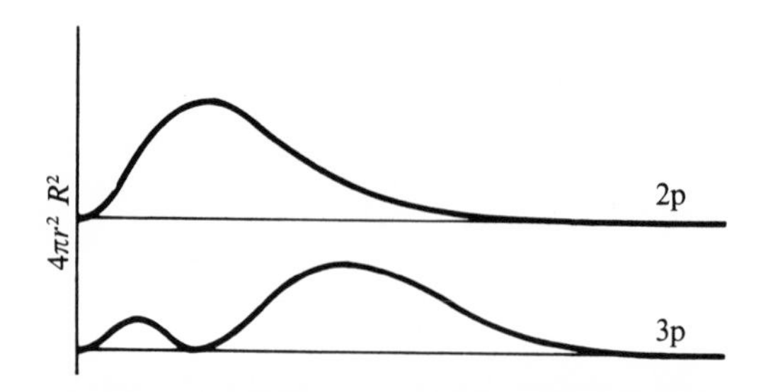

FIG. 7.6. Probability of finding the electron at the distance r from the nucleus, in the first two p functions of the hydrogenlike atom.

Fig. 7.6 shows the radial dependence of the probability functions in the same two p states, for spherical shells of radius r, to be compared with the s-state functions in Fig. 7.4. Again the most probable distance of the electron from the nucleus increases as n increases. That peak probability is at a slightly smaller value of r in a p state than in the s state for the same value of n. But this difference between s states and p states cannot be given a ready interpretation because of the complication introduced in p states by their angular dependence.

APPENDIX

RADIAL DEPENDENCE OF HYDROGENLIKE STATES

The wave functions for an electron attracted to a single nucleus can be expressed in the form $\psi = RL$, where R is a function of r, the distance of the electron from the nucleus, and L is a function of the direction of the vector $\mathbf{r}$, as Chapter 5 described. The radial dependence R must then obey the ordinary differential equation:

$$\frac{\mathrm{d}^2(rR)}{\mathrm{d}r^2}+\left[E-\frac{l(l+1)}{r^2}+\frac{2Z}{r}\right](rR) = 0, \tag{A7.1}$$

and acceptable solutions to this equation must be quadratically integrable over all space.†

Notice in the first place that at very large r (A7.1) approaches

$$\frac{\mathrm{d}^2(rR)}{\mathrm{d}r^2}+E(rR) = 0. \tag{A7.2}$$

Since our interest is in bound states, where E is negative, we can expect from (A7.2) that the solutions will look roughly like $\mathrm{e}^{-\alpha r}$, where $\alpha^2 = -E$, at large r. This suggests examining solutions having this factor, and it turns out to be convenient to take out a factor r^{l+1} at the same time. Substituting $rR = \mathrm{e}^{-\alpha r}r^{l+1}P$ in (A7.1) yields a differential equation for P:

$$\frac{\mathrm{d}^2P}{\mathrm{d}r^2}+2\left(\frac{l+1}{r}-\alpha\right)\frac{\mathrm{d}P}{\mathrm{d}r}+\frac{2}{r}[Z-\alpha(l+1)]P = 0. \tag{A7.3}$$

It is now appropriate to seek a solution for P in the form of a power series‡ in r:

$$\begin{aligned} P &= a_0+a_1r+a_2r^2+\ldots a_jr^j+\ldots \\ \frac{\mathrm{d}P}{\mathrm{d}r} &= a_1+2a_2r+3a_3r^2+\ldots+(j+1)a_{j+1}r^j+\ldots \\ \frac{\mathrm{d}^2P}{\mathrm{d}r^2} &= 2a_2+2\,.\,3a_3r+3\,.\,4a_4r^2+\ldots+(j+1)(j+2)a_{j+2}r^j+\ldots\,. \end{aligned} \tag{A7.4}$$

Substituting (A7.4) into (A7.3), and requiring that the equation should be identically satisfied in each of the powers of r, provides relations that must be satisfied between

† This requirement is discussed in *Stationary states*, Chapter 2.

‡ An analogous procedure for obtaining the wave functions of a one-dimensional harmonic oscillator is discussed in *Stationary states*, Chapter 5.

the coefficients in the power series:

$$\begin{aligned} a_1 &= -\frac{Z-\alpha(l+1)}{l+1}a_0, \\ a_2 &= -\frac{Z-\alpha(l+2)}{2l+3}a_1, \\ a_j &= \frac{-2}{j}\frac{Z-\alpha(l+j)}{2l+j+1}a_{j-1}. \end{aligned} \tag{A7.5}$$

These are *recursion relations*, which permit a determination of the coefficients a_j in succession, starting with some assumed value of a_0, which remains as the one arbitrary constant multiplying the solution. In general the series of powers of r in P will be an infinite series.

Looking at what happens to a_j as j gets very large reveals the appearance of the series 'far to the right'—the relation between the coefficients of large powers of r in the series. The large value of j then dominates everything else in the recursion relations (A7.5), which become

$$a_j = \frac{2\alpha}{j}a_{j-1}.$$

This is the same as the relation between successive coefficients in the series

$$e^{2\alpha r} = 1+2\alpha r+\frac{(2\alpha r)^2}{2!}+\ldots+\frac{(2\alpha r)^{j-1}}{(j-1)!}+\frac{(2\alpha r)^j}{j!}+\ldots\,.$$

Hence the solution for P will diverge as $e^{2\alpha r}$ for large r, yielding $R = r^l e^{+\alpha r}$, an inadmissible wave function for a bound state.

The only salvation is to choose the quantities in (A7.5) in such a way as to make the series terminate, so that P is a polynomial in r rather than an infinite series. Such a termination can be accomplished by taking $Z = \alpha(l+j)$. Then a_j and all succeeding coefficients vanish:

$$P = a_0+a_1r+\ldots+a_{j-1}r^{j-1},$$

and thus

$$R = e^{-\alpha r}r^l(a_0+a_1r+\ldots+a_{j-1}r^{j-1}). \tag{A7.6}$$

Since j, l, and Z are whole numbers, the condition is

$$\alpha^2 = -E = \frac{Z^2}{n^2}, \tag{A7.7}$$

where n is the whole number $n = l+j$. The polynomials defined in this fashion are called the *Laguerre polynomials.*

Eqn (A7.7) then determines the permitted energy levels for an electron in this sort of well. Notice, from (A7.6), that n cannot be less than $l+1$, for there must be at least one term in the polynomial. Thus the permitted values of energy are

$$-E = Z^2/(l+1)^2, \qquad Z^2/(l+2)^2, \text{ etc.}$$

The lowest energy level of all—the level E_0 with the largest negative energy—will be that in which $l = 0$, $n = 1$, and thus $-E_0 = Z^2$ in atomic units. If there is only one electron in the atom, its wave function will normally be that corresponding to this lowest level: the *ground state* $\psi_0 = R_0L_0 = Ae^{-Zr}$.

PROBLEMS

7.1 By using the quantum condition (A7.7) and the recursion relation (A7.5), derive the radial parts (A7.6) of the wave functions:

$$1s:\ R \simeq e^{-Zr}$$

$$2s:\ R \simeq e^{-Zr/2}\left(1-\frac{Zr}{2}\right)$$

$$3s:\ R \simeq e^{-Zr/3}\left(1-\frac{2Z}{3}r+\frac{2Z^2}{27}r^2\right)$$

$$2p:\ R \simeq e^{-Zr/2}r$$

$$3p:\ R \simeq e^{-Zr/3}r\left(1-\frac{Z}{6}r\right)$$

$$3d:\ R \simeq e^{-Zr/3}r^2.$$

7.2 Derive the normalizing coefficient for the complete 1s wave function, in which all space is accessible to the particle.

7.3 Calculate the mean potential energy of an electron in the state described by a 1s wave function, and compare it with the total energy.

7.4 Calculate the mean distance of the electron from the proton in a hydrogen atom in the 1s state.

7.5 Show that the *most probable* distance from the nucleus at which you will find an electron, when it is in the 1s wave function of a hydrogenlike atom, is $1/Z$ atomic units. It is this value, not that of Problem 7.4, that is most appropriately taken as the 'radius' of the electronic 'orbit'.

7.6 (a) Show that choosing the atomic unit of energy twice as large as the unit used here (Discussion 4.1) has the *aesthetic advantage* of corresponding with a choice of units in which the mass of the electron is one, the charge on the electron is one, and the unit of action $h/2\pi$ is one.
(b) Show that the choice (a) has the *aesthetic disadvantage* of introducing a factor of two at one point in Schrödinger's equation for the electron in a central Coulomb potential.

8. The Building-up Principle

AN early and major triumph of the wave-mechanical theory of electronic behaviour in atoms was its explanation of the periodic table of the elements.† In this application of the theory the results obtained in the last three chapters, gratifyingly exact for atoms and ions with only one electron, are extended qualitatively to atoms with more than one electron. For that extension an atom is regarded as a system of identifiable electrons, each occupying a state similar to one or another of the one-electron states described in the last chapter. Limitations are placed upon their occupancy of these states by using the exclusion principle‡ in its simplest form: at most two electrons, one with spin 'up' and the other with spin 'down', can occupy any one of the states.

The most important aspects of the periodic table can be explained by applying these principles to describe the *ground states* of the different species of atoms—the states of lowest energy. The application can be systematically carried out by imagining, for an atomic species whose atomic number is Z, that a bare atomic nucleus of charge Ze acquires its Z electrons one by one. This procedure is sometimes called the 'building-up principle'—a necessarily awkward translation of the German *aufbauprinzip.*

The last chapter has already described what happens when the first electron is associated with the nucleus of charge Ze. The electron finds a choice of possible states similar to those for the electron in hydrogen. The only difference is that the wave functions are pulled in closer to the nucleus, because its attractive charge is Ze instead of e. In the unexcited atom, the electron will occupy the state of lowest energy, the 1s state.

When a second electron is added, it will try to occupy the same state, aligning its spin antiparallel with the spin of the first electron, as the exclusion principle describes. Now neither electron will experience simply the potential $-(Ze^2/4\pi\epsilon_0 r)$, because in addition each will repel the other. Since the 1s wave function gives an especially high probability for finding both electrons in a small region near the nucleus (Fig. 7.3), their mutual repulsion will tend to drive them apart and thus to modify their wave functions appreciably.

The atom with $Z = 2$ is helium, a species with great stability and chemical inertness. The energy that must be supplied to remove one of the two electrons of a helium atom is 1·81 times that required to remove the one electron from a hydrogen atom. If the two electrons were not repelling each other, the removal of one would require four times the energy required for hydrogen, since the energy associated with the hydrogenlike state varies as Z^2, as the

† See the tables at the back of this book.

‡ The exclusion principle for electrons is discussed in *Stationary states*, Chapter 9.

last chapter showed. On the other hand, if the first electron counteracted half the attraction of the nucleus, the ionization energy would be the same as that of hydrogen. The next two chapters discuss more quantitatively some ways of predicting where the true ionization energy will fall between these limits.

When the nuclear charge is raised to $3e$, and a third electron is added, that electron cannot adopt the 1s state because that state is 'fully occupied'. According to the exclusion principle, no two electrons can occupy states with the same four quantum numbers, and the third electron finds two electrons already in the 1s state (and thus with the same values of n, l, and m), each with one of the two possible spin quantum numbers, $m_s = \pm\frac{1}{2}$. The only recourse for the third electron is to adopt the hydrogenlike state of next higher energy.

The treatment of the hydrogenlike states in the last chapter may seem to leave no clue to what that state will be. The states next higher in energy are those with $n = 2$ and they are degenerate: four, with different values of l and m, have the same energy. In fact, however, it is not hard to see that, in an atom with $Z = 3$, the 2s state will have lower energy than the 2p states. The two electrons occupying the 1s state form a spherical cloud of negative charge. In so far as the third electron stays outside that cloud, it sees an attractive charge of only $+e$ rather than $+3e$. The forms of the s and p wave functions (Figs. 7.3, 7.4, 7.5, and 7.6) show that s electrons will penetrate that cloud more deeply than p electrons, and thus will experience more attraction from the nucleus. The third electron will therefore adopt a 2s state; so also would a fourth, with opposite spin.

The element of atomic number 3 is lithium, the lightest of the alkali metals. Its ionization potential is less than forty per cent of that of hydrogen. Its readiness to form ionic salts, in which it exhibits unit valency, follows from the fact that its one 'outer' electron sees on the average an apparent nuclear charge much less than $+3e$, and moreover is further from the centre of that attractive charge than the single 1s electron in hydrogen.

It might seem natural to proceed further along the periodic table by using the reasoning so far employed, in the following way. Each state, derived from thinking of an atom as 'hydrogenlike', can accommodate at most two electrons. When all the states belonging to a value of the principal quantum number n have been filled, the next electron should adopt a state for which n is one unit larger. Since there are n^2 states of principal quantum number n, a *rare gas* atom should appear at atomic numbers $2\times 1^2 = 2$, $2\times 1^2+2\times 2^2 = 10$, $2\times 1^2+2\times 2^2+2\times 3^2 = 28$, etc.

This simple reasoning is satisfactory for helium and neon, but it breaks down beyond that, for a reason that becomes clear on examining further the repulsions offered by the growing cloud of electrons. The difference in penetration of the states of different l makes the relative energies of the states of any one n go up with l, when the actual occupation of the inner state by electrons is taken into account. In consequence, the approximate stability sequence for filling the various states can be illustrated as in Fig. 8.1. There each state is

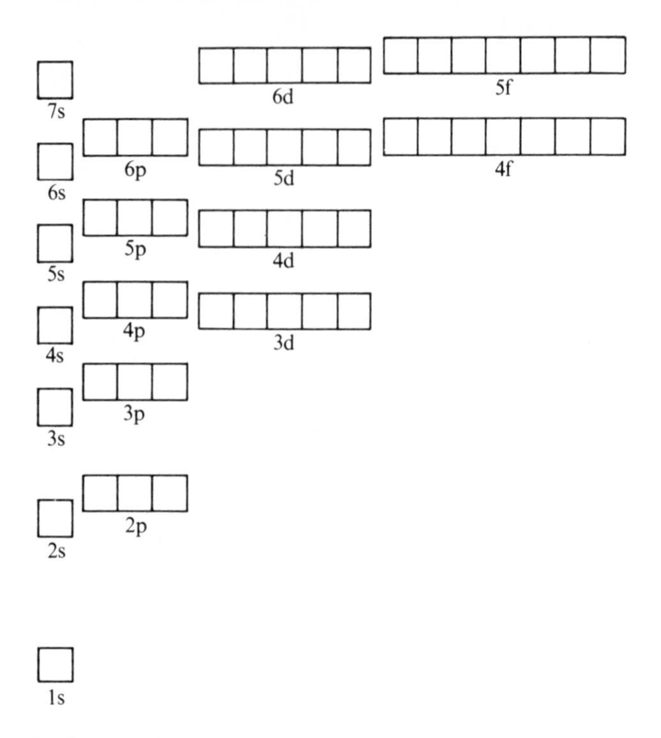

FIG. 8.1. A rough diagram of the sequence of energies of the one-electron states in a many-electron atom. Each state can accommodate two electrons, and with increasing atomic number the states of higher energy are occupied successively.

represented by a slot, which can be occupied by two electrons; by way of example, the occupancy in scandium ($Z = 21$) is shown in Fig. 8.2. A rare gas, argon, is encountered at $Z = 18$ rather than $Z = 28$ because the 3d states penetrate toward the nucleus so little that their energies are much higher than those of the 3p states—even higher than that of the 4s state, despite its higher principal quantum number.

Unfortunately no diagram like Fig. 8.1 can be drawn that is completely satisfactory, because the successive filling of the slots 'weighs them down', so to speak. In the upper part of the diagram, where levels differ less in energy, that weighing down sometimes inverts the order of stability, and thus changes the order of filling. The elements of the top row of Period 5 of the periodic table furnish a good illustration of these changes. Proceeding from the alkali metal rubidium ($Z = 37$) through zirconium ($Z = 40$) two electrons are placed in the 5s slot, then two in the 4d slots as Fig. 8.1 would suggest. But now the 4d slots have fallen to about the level of the 5s slot: in an atom of the next element, niobium ($Z = 41$), there is only one 5s electron and there are four 4d electrons. Thenceforward the filling of the 4d slots continues, retaining only one 5s electron, until palladium ($Z = 46$) is reached. There no 5s electron appears, and abruptly all the five 4d slots are filled with ten electrons. Accordingly in the next element, silver, occupancy of the 5s slot begins again; it has ten 4d electrons and one 5s electron, not nine 4d electrons and two 5s electrons as Fig. 8.1 would suggest.

The fact that silver has one 5s electron is responsible for the chemical behaviour that places it in Group I of the periodic table. Characteristically it exhibits the valency one, in such familiar compounds as silver chloride (AgCl). But silver does form compounds with very avid electron acceptors—compounds such as silver difluoride (AgF_2)—in which it exhibits the valency two, much as if it had two 5s electrons.

The scheme by which one-electron levels are filled in a many-electron atom is often shown by a notation in which the successive levels are listed, and the number of electrons in each sort is exhibited as a superscript. Thus the notation for the ground state of scandium (Fig. 8.2) would be $1s^2$ $2s^2$ $2p^6$ $3s^2$ $3p^6$ $3d^1$ $4s^2$. This is usually abbreviated to $3d^1 4s^2$, mentioning only the occupancy of the uppermost horizontal belt in Fig. 8.1 that is occupied at all, in other words the levels that are open to doubt.

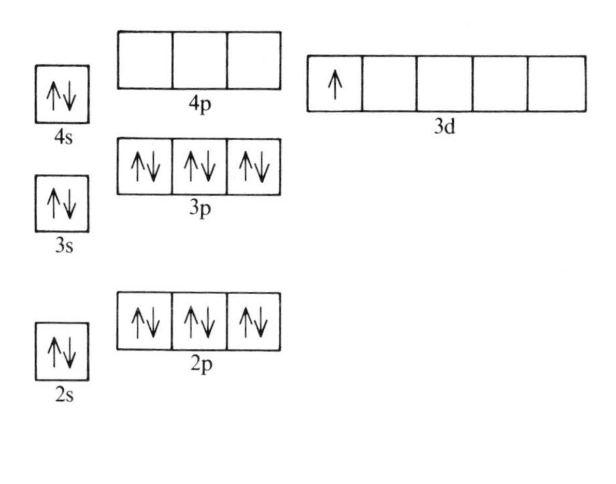

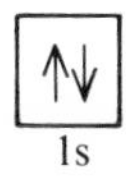

FIG. 8.2. Scandium illustrates the fact that the 3d levels are at higher energy than the 4s in the ground states of most of the atoms of Table 2.4, period 4.

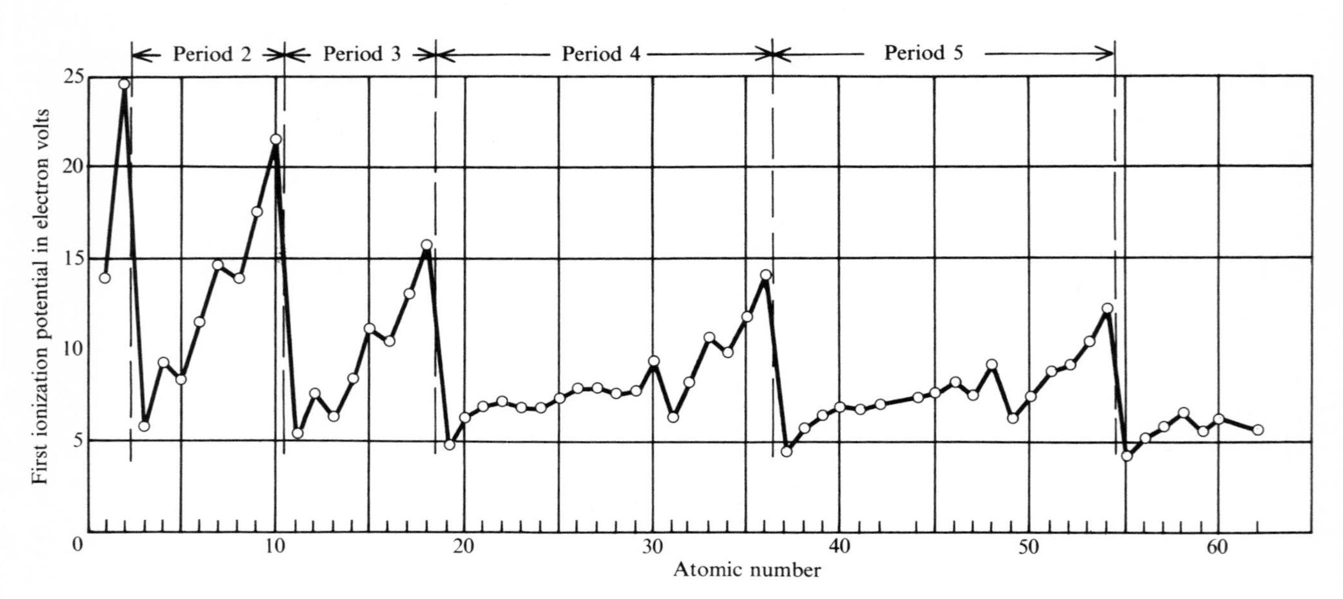

Fig. 8.3. The first ionization energies of the elements, plotted in electron volts against atomic number.

So far the reasoning has left open the question of the order in which such slots as the 2p slots are filled—those that are shown in groups on the same level in Fig. 8.1. In the hydrogenlike atom, these states are degenerate: they have the same energy, as Chapter 7 showed. But they are distinguishable in the angular distribution of their probability clouds, as appeared in Chapter 5. From examining Fig. 5.3 it is clear that, when one of these states is occupied by an electron, a second electron will prefer to occupy a different one of these states because that will enable the two electrons to stay further apart under their mutual electrostatic repulsion. In consequence, the states of any such group tend to be filled singly, so far as possible. Double occupancy of a slot begins only after each slot in such a group has already received one electron.

Some of the chemical properties of the atomic species summarized in the periodic table receive a ready explanation in the terms used in this chapter. Thus the halogens (F, Cl, Br, and I) have a 'negative' valency of one because they can accept just one more electron without a serious increase in energy. With the extra electron, their electronic configurations become those of the rare gases, but with a net charge of $-e$ unbalanced by the charge on their nuclei. Hence their ions have the spherical symmetry of the rare gas configurations (see Problem 5.1).

Similarly an atom of an alkali is a ready donor of one electron, acquiring a spherical rare-gas configuration with net charge $+e$. In Fig. 8.3 the *first ionization potentials* of the elements—the energies required to remove one electron from an atom—are plotted in electron volts, to show that this physical property exhibits the periodicity of the periodic table. The rare gases provide the maxima, the alkali metals the minima. It is not surprising that a halogen atom can capture an electron from an alkali metal atom, and thus that the resulting ions—the positive alkali ion and the negative halide ion—attract each other and form an *ionic bond*.†

Often an atom engages in bonding by sharing electrons with its partner, to form a *covalent bond*. During the fraction of time that its partner's electron is near it, that electron will appear to occupy the lowest energy state that its electron configuration affords. And while its electron is near its partner, that electron will seek the lowest state the partner affords. Indeed the covalent bond is characteristically an 'electron-pair' bond. Two electrons with opposite spins occupy a state whose wave function looks like a combination of the lowest available states in each atom, connected together between them. Such bonds will form only when there are states available in the participating atoms to accommodate the additional shared electrons without exorbitant increases in energy.

Two fluorine atoms, for example, will bond together to form a fluorine molecule, with each atom contributing an electron to the bond. Here each

† The physical character of these and other bonds mentioned in this chapter are discussed quantitatively in *Bonds between atoms*.

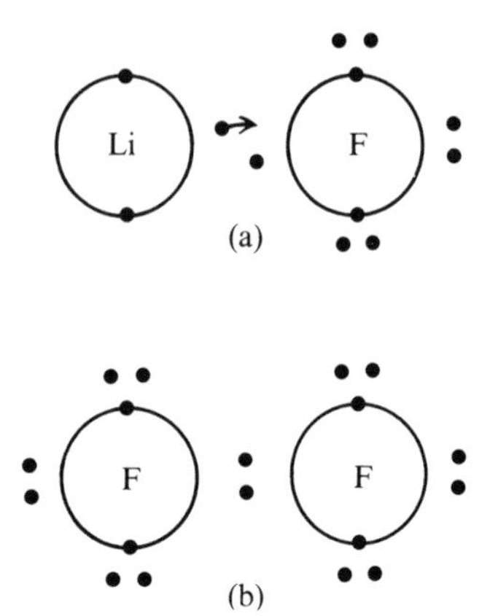

FIG. 8.4. Schematic illustration of the way electrons participate in bonding and give an element a characteristic valency: (a) the ionic lithium fluoride, (b) the covalent fluorine molecule.

atom offers one unoccupied 2p state for its neighbour's electron. Two neon atoms, on the other hand, could accommodate each other's electrons only in 3s states, and there is no reason for an electron to stray from the 2p state, which it occupies in one neon atom, into the 3s state, a state of much higher energy, offered by another neon atom. Hence neon atoms do not form chemical bonds with one another.

Fig. 8.4 shows a rough way of symbolizing what is going on in bonding. At (a) an atom of lithium loses its one 2s electron to an atom of fluorine, completing the shell of eight electrons of principal quantum number 2 in the fluorine atom, and forming the ionic bond in lithium fluoride. At (b) two fluorine atoms share a pair of electrons, to form a diatomic molecule

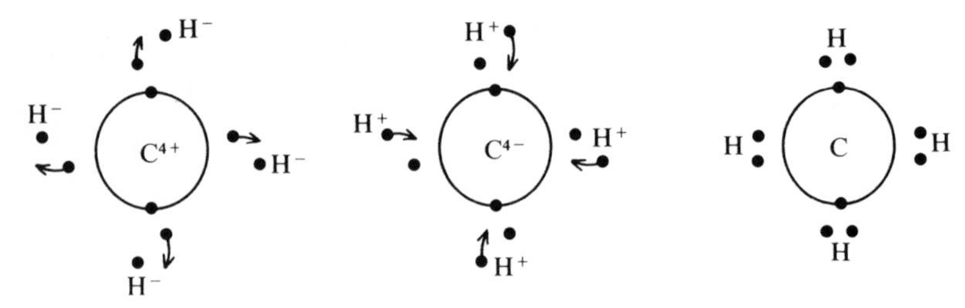

FIG. 8.5. Of three ways for schematizing the bonding in the methane molecule, CH_4, the covalent scheme at the right is most probable, because the substance shows no evidence of the ionic character implied by the schemes that give to the central carbon atom a large net charge.

covalently bonded, in which every state with $n = 2$ in both atoms is occupied by two electrons.

Fig. 8.5 shows three alternatives that might be written for a molecule of methane, CH_4. In one, a carbon atom is stripped of its outer electrons, becoming an ion of charge $4e$ with the configuration of helium, and bonding ionically with the four hydrogen atoms by filling their 1s states and giving them also the helium configuration. Another illustrates the reverse hypothesis: the carbon atom acquires a charge $-4e$ and the configuration of neon, bonding ionically to naked protons. In the third, covalent bonds are formed between the hydrogen atoms and the carbon atom, keeping each neutral, and filling the 1s states of the former, and all the states with $n = 2$ in the latter. The last hypothesis is certainly the best, because the large net atomic charges invoked by the two other hypotheses would surely push and pull the electrons back where they came from. Notice, however, that the valency of carbon is four, by any of these three arguments.

PROBLEMS

8.1 Use the stability diagram of Fig. 8.1 to answer the questions.
(a) What levels are being filled as you traverse the elements from scandium to zinc in the periodic table?
(b) What levels are being filled as you traverse the *rare earths*?

8.2 Would you expect the densities of the solid elements to reflect in any way the periodicity of the periodic table?

8.3 Give the full and the abbreviated notation for the extent of filling of the one-electron levels in the carbon atom in its ground state.

8.4 The argument of the text regarding the 4d *transition elements* and silver suggests a similar argument regarding the 3d *transition elements* and copper. Here the initial difference between the 3d and 4s levels is greater than that between the 4d and 5s levels, and the 'weighing-down' effect does not succeed in equalizing them, with the interesting exception of chromium (notice that consequently the 3d levels are precisely half filled in Cr), until you reach the end of the period. Thus the ground state of the nickel atom is $3d^8 4s^2$ and that of copper is $3d^{10} 4s^1$. In view of this, discuss the fact that the most prominent valency state of copper is two (though it does exhibit the valency one), whereas the most prominent valency state of silver is one (though it does exhibit the valency two).

9. The Two-electron Delta-well Atom

THE success of Chapter 7 in calculating exactly the properties of the hydrogen-like atom may spur you to undertake exact calculations for an atom with more than one electron. It turns out that, at present, only approximate calculations are feasible. Perhaps you will accept the hopelessness of exact calculation with better grace after looking at the difficulties that even a simplified model presents. In fact such an examination offers more than that negative virtue: it exemplifies in a simple context many of the principles that must govern any exact thinking about electronic behaviour.

A very simple-looking model of a two-electron atom can be constructed by extending the one-dimensional delta-well model of Chapter 4 to include two electrons instead of one. Its analysis illustrates the sort of thinking that must be applied to a wave function of several variables that are not coordinates along the three dimensions of real space. Furthermore, an approximate calculation of the energy of the model by the variational method† agrees remarkably well with the true energy of the helium atom. The procedure is a simplified version of that which has been used to make the most accurate calculations of the energy of the helium atom.

Think first of two electrons, both free to move in only one dimension x, and a delta well at $x = 0$ representing the attractive potential due to the nucleus. The wave functions for the system are now functions of two variables—the coordinates x_1 and x_2 of the two electrons—and Schrödinger's equation (written in atomic units) is the partial differential equation in two variables:

$$\frac{\partial^2\psi}{\partial x_1^2}+\frac{\partial^2\psi}{\partial x_2^2}+[E-V(x_1, x_2)]\psi = 0. \tag{9.1}$$

The potential $V(x_1, x_2)$ now resembles Fig. 9.1: whenever $x_1 = 0$, electron No. 1 is at the potential well of depth V and width l; whenever $x_2 = 0$, electron No. 2 is at the well. Looking down on the representation in Fig. 9.1, and letting $V \doteq \infty$ and $l \doteq 0$ in such a way that $Vl = \eta$, we can represent the potential by Fig. 9.2.

So long as the problem remains in this form, it offers no difficulty, but it offers no novelty. The two electrons are quite independent; neither is affected by its companion. The potential can be written $V(x_1, x_2) = V(x_1)+V(x_2)$,

† The variational method of approximating the energy of the ground state of a wave-mechanical system is discussed in *Stationary states*, Chapter 7.

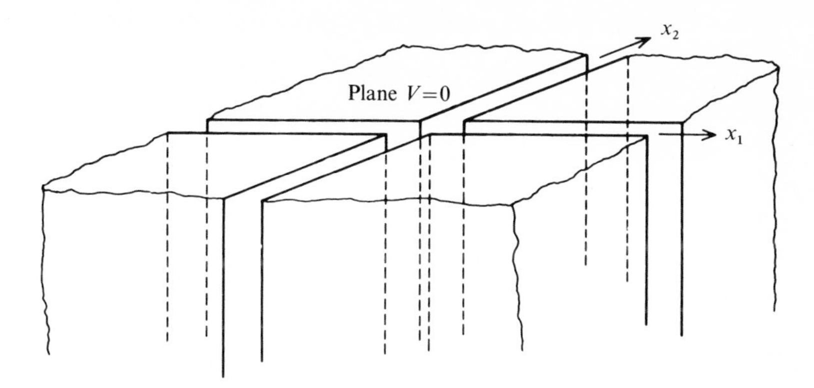

FIG. 9.1. Delta wells of potential energy for two electrons free to move in only one dimension of real space.

and the wave function can be written $\psi(x_1, x_2) = \psi_1(x_1)\,.\,\psi_2(x_2)$. Substituting these into (9.1) yields

$$\psi_1 \frac{d^2\psi_2}{dx_2^2} + \psi_2 \frac{d^2\psi_1}{dx_1^2} + [E - V(x_1) - V(x_2)]\psi_1\psi_2 = 0. \qquad (9.2)$$

Divided by $\psi_1\psi_2$, (9.2) becomes

$$\frac{1}{\psi_2}\frac{d^2\psi_2}{dx_2^2} - V(x_2) + \frac{1}{\psi_1}\frac{d^2\psi_2}{dx_1^2} - V(x_1) = -E. \qquad (9.3)$$

Here the variables are separable: you can take

$$\frac{1}{\psi_2}\frac{d^2\psi_2}{dx_2^2} - V(x_2) = -E_2,$$

$$\frac{1}{\psi_1}\frac{d^2\psi_2}{dx_1^2} - V(x_1) = -E_1, \qquad E_1 + E_2 = E. \qquad (9.4)$$

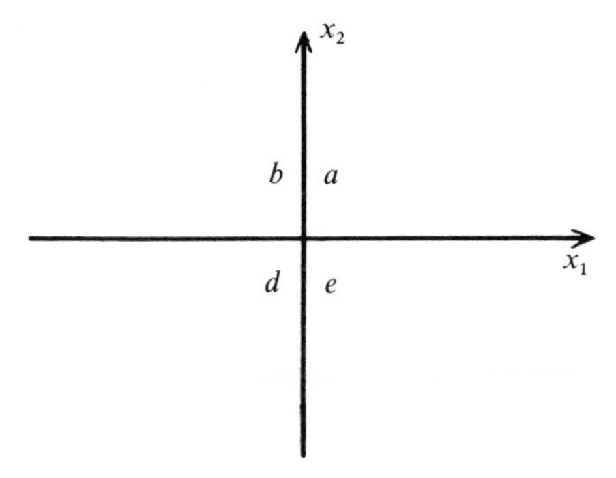

FIG. 9.2. Plan view of the delta wells of Fig. 9.1.

Each differential equation in (9.4) applies to one electron alone with energy E_1 or E_2. The energy of the whole system is the sum of the two one-electron energies, and the wave function is the product of the two one-electron wave functions. For the four quadrants of Fig. 9.2, the wave function for the pair of electrons can be written

$$\psi_a = Ae^{-\alpha(x_1+x_2)}, \qquad \psi_b = Ae^{-\alpha(x_2-x_1)},$$
$$\psi_d = Ae^{\alpha(x_1+x_2)}, \qquad \psi_e = Ae^{-\alpha(x_1-x_2)}, \tag{9.5}$$

where $2\alpha^2+E = 0$.

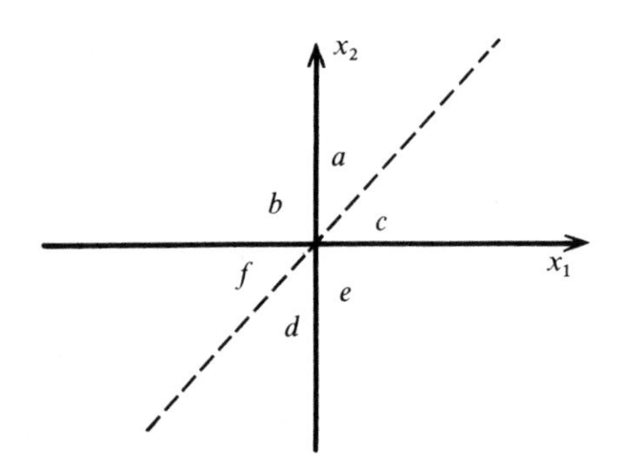

FIG. 9.3. A delta ridge along the dotted line simulates the mutual repulsion of the two electrons.

In the real physical system of interest, however, the two electrons repel each other according to Coulomb's law, just as the nucleus attracts both electrons by that law. Since the model simulates the Coulomb attraction by a delta well, it is appropriate to simulate the Coulomb repulsion by a *delta ridge*. The system encounters that ridge when the two electrons have the same coordinate: when $x_1 = x_2$. Then Fig. 9.2 must be modified as in Fig. 9.3, where the dotted line is a plan view of the ridge.

You can predict confidently a few qualitative features of the wave function for this problem. Everywhere except at the ridge and the well it will be a solution of the wave equation for free electrons, in other words, with $V = 0$. At the ridge and the well the function will be continuous, but it will have a discontinuity of slope, of amount determined by the product $Vl = \eta$ times the value of the wave function there. Previous work in Chapter 4 with the one-electron case shows that, when there is no potential ridge (Fig. 9.2), the wave function (9.5) will look like the roof of a pagoda (Fig. 9.4), with four sections joining at ridge poles and dropping off from them exponentially.

When the potential ridge is introduced, two of the four sections of the roof will acquire V-shaped gutters: the discontinuity of slope at the bottom of the gutter will be similar to that at the well but of opposite sign. The reduced values of the wave function at the gutters show that the probability of finding the two electrons close together is reduced by their mutual repulsion. Since some probability will be pushed away from the dotted line over its entire course, the roof will be lower at the centre and will slope away from the centre less steeply. And the energy will be higher (less negative) because the repulsion between the electrons will increase their potential energy at all separations.

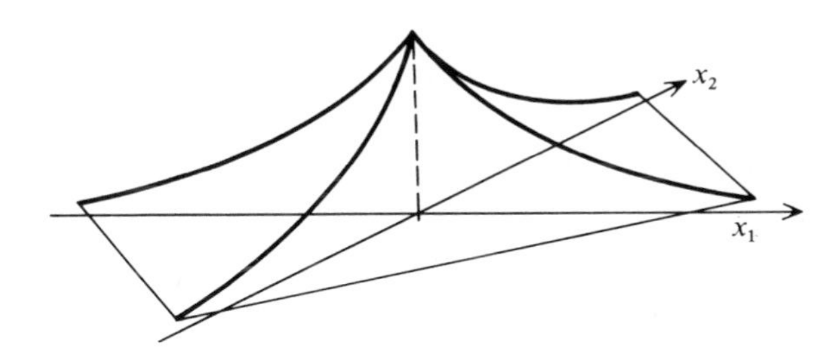

FIG. 9.4. The wave function (9.5) for two electrons moving in one dimension in the presence of a delta well, when their mutual repulsion is ignored.

The quantitative task is to construct solutions appropriate to the six regions separately lettered in Fig. 9.3, to match them at the boundaries of those regions, and to evaluate the discontinuities of slope at those boundaries. The magnitude of the task can be reduced at the outset by observing that the problem exhibits certain symmetries. Since the electrons are identical, then (if their spins are opposite) the whole wave function must be symmetric to interchange of x_1 and x_2. In other words

$$\begin{aligned} \psi_c(x_1, x_2) &= \psi_a(x_2, x_1), \\ \psi_e(x_1, x_2) &= \psi_b(x_2, x_1), \\ \psi_f(x_1, x_2) &= \psi_d(x_2, x_1). \end{aligned} \tag{9.6}$$

Moreover, the behaviour of the electrons on one side of the well is the same as their behaviour on the other side, and thus the whole wave function must be symmetric or antisymmetric to the act of transferring them from side to side. Since Chapter 4 has shown that the delta well provides no antisymmetric bound state for one electron, it is permissible to guess that the well will not provide such a state here, for the energy of all states is increased here by adding a repulsive potential. Choosing the symmetrical state, already visualized in Fig. 9.4 but with gutters added, this second symmetry requirement can be embodied in the wave function by writing

$$\begin{aligned}\psi_d(x_1, x_2) &= \psi_a(-x_1, -x_2),\\ \psi_c(x_1, x_2) &= \psi_f(-x_1, -x_2),\\ \psi_e(x_1, x_2) &= \psi_b(-x_1, -x_2).\end{aligned} \tag{9.7}$$

Notice that the relations (9.6) and (9.7) reduce to ψ_a and ψ_b the separate forms of wave functions that must be examined: the others can be constructed

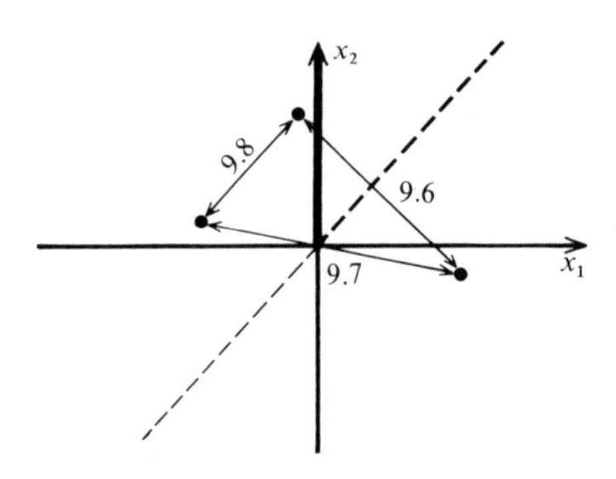

FIG. 9.5. The symmetry relations (9.6) and (9.7) imply the symmetry relation (9.8).

from those two. Furthermore, the two relations between ψ_e and ψ_b imply a symmetry in ψ_b itself:

$$\psi_b(x_1, x_2) = \psi_b(-x_2, -x_1). \tag{9.8}$$

Fig. 9.5 shows the relationship between three typical points in the (x_1, x_2) plane to which the symmetry relations for ψ_e and ψ_b apply. In consequence, the boundaries at which the wave function and its derivatives must be evaluated are reduced to the two marked in heavy lines in Fig. 9.5; the symmetry relations automatically take care of the rest.

Now consider how to make some modification of the wave function (9.5) that might suit it to the new conditions. Taking $\psi_b = Ae^{(\alpha_1 x_1 - \alpha_2 x_2)}$, where $\alpha_1^2+\alpha_2^2+E = 0$, provides a solution to the wave equation for the free electrons with negative energy. The signs of α_1 and α_2 must be positive in the region b in order to make the solution approach zero with increasing negative values of x_1 and positive values of x_2 in such a way that the integral of its square over the quadrant b is finite. Suitable also would be $\psi_c = Ae^{(\alpha_2 x_1 - \alpha_1 x_2)}$; and the sum of the two,

$$\psi_b = A[e^{(\alpha_1 x_1 - \alpha_2 x_2)} + e^{(\alpha_2 x_1 - \alpha_1 x_2)}], \tag{9.9}$$

is a solution that obeys the symmetry requirement (9.8).

In the region a there is no symmetry requirement: for that region $A_1 e^{-(\alpha_1 x_1 + \alpha_2 x_2)}$ and $A_2 e^{-(\alpha_2 x_1 + \alpha_1 x_2)}$ can be added with different coefficients A_1 and A_2. And a solution $A_3 e^{(\alpha_1 x_1 - \alpha_2 x_2)}$ can be added if α_1 is made less than α_2, since x_1 is always less than x_2 in the region a and this solution will then be quadratically integrable over the octant a. Clearly this term will help to produce a gutter at $x_1 = x_2$, for if A_3 is negative the term will fall off increasingly as x_1 increases toward x_2. In summary, a suitable choice is

$$\psi_a = A_1 e^{-(\alpha_1 x_1 + \alpha_2 x_2)} + A_2 e^{-(\alpha_2 x_1 + \alpha_1 x_2)} + A_3 e^{(\alpha_1 x_1 - \alpha_2 x_2)}, \tag{9.10}$$

and immediately the symmetry requirement (9.6) provides

$$\psi_c = A_1 e^{-(\alpha_2 x_1 + \alpha_1 x_2)} + A_2 e^{-(\alpha_2 x_2 + \alpha_1 x_1)} + A_3 e^{(\alpha_1 x_2 - \alpha_2 x_1)}. \tag{9.11}$$

Now ψ_b and ψ_a must be matched at $x_1 = 0$, and ψ_c and ψ_a must be matched at $x_1 = x_2$. The second requirement is automatically satisfied: the method of constructing ψ_c from ψ_a ensures that they are equal at $x_1 = x_2$. To satisfy the first requirement, notice that $\psi_b = \psi_a$ at $x_1 = 0$ if

$$A_2 = A, \quad \text{and} \quad A_1 + A_3 = A. \tag{9.12}$$

It remains to impose the requirements that the first derivatives of the wave function should exhibit the discontinuities invoked by the well and by the ridge.

Giving the derivative of the wave function the proper discontinuity at the well requires (cf. (D4.2))

$$\left.\frac{\partial \psi_a}{\partial x_1}\right|_{x_1=0} - \left.\frac{\partial \psi_b}{\partial x_1}\right|_{x_1=0} = -\eta_n \psi|_{x_1=0}. \tag{9.13}$$

Here η_n (read 'eta at the nucleus') is a positive number specifying the well: the limit of the product Vl of the depth of the well by its width. The requirements implied by (9.13) can be written

$$A_2 = \left(\frac{\eta_n}{\alpha_1} - 1\right) A, \quad \text{and} \quad A_1 - A_3 = \left(\frac{\eta_n}{\alpha_1} - 1\right) A. \tag{9.14}$$

Combining these with (9.12) yields

$$\begin{aligned} 2\alpha_2 &= \eta_n, & A_1 &= \frac{\alpha_2}{\alpha_1} A, \\ A_3 &= \left(1 - \frac{\alpha_2}{\alpha_1}\right) A, & A_2 &= A. \end{aligned} \tag{9.15}$$

Only one quantity, α_1, remains free for adjusting the wave function so that its first derivative will exhibit a suitable discontinuity at the ridge.

To set up the requirement on that discontinuity at $x_1 = x_2$, it is simplest to rotate the coordinates so that one of them lies along the ridge. The formalities of this process are outlined in Discussion 9.1. It turns out that the requirement can be met only by satisfying two more relations between the A's and the α's, analogous to (9.14), which cannot be accomplished with the one remaining free constant α_1.

This failure implies that the exact solution to the problem cannot be expressed in terms of the simple exponential solutions to the wave equation with which this analysis started. Instead of seeking an exact solution in less familiar mathematical terms, it is more appropriate to obtain an approximate solution that retains the virtues of the work already accomplished. Surely the general form of the wave function (9.9) and (9.10) has many of the properties that the true wave function must have. Supposing that the true function is somewhat like it, you can use this form to calculate approximations to certain properties of the system, in particular its energy.

Since Discussion 9.1 shows that this form cannot be an eigenfunction of the energy, the relation $\alpha_1^2+\alpha_2^2+E = 0$ must be discarded. But an expectation† value for the energy associated with any such wave function can be calculated by evaluating

$$\bar{H} = \frac{\int \psi^* H \psi \, d\tau}{\int \psi^* \psi \, d\tau}. \tag{9.16}$$

Moreover the variation theorem asserts that the minimum value of $\bar{H}$ with respect to all the free parameters in (9.16) will be an upper limit to the true energy of the system in its lowest state.

How many independent parameters are accessible in the proposed form? The relations (9.12) ensure the continuity of the function, and thus only one of the A ratios is free. If you retain the requirement that the discontinuity of derivatives at the wells shall be precisely right, the relations (9.15) remove the freedom of the A's (except of course the freedom to choose one as the normalizing constant), and pin down α_2, leaving α_1 free. But it is aesthetically unpleasant to single out wells for favoured treatment, in contrast to ridges. Hence a suitable approximate wave function might have three independent parameters, α_1, α_2, and an A ratio, and the quantity $\bar{H}$, calculated by (9.16), would be minimized with respect to all three. A convenient way to retain the single degree of freedom in the A's would be to modify (9.15) to read $A_1 = \lambda A$, $A_3 = (1-\lambda)A$, and take λ as the free A parameter.

It will reduce algebraic labour, however, to adopt a compromise: leave α_1 and α_2 free, but take $\lambda = \alpha_2/\alpha_1$ as in (9.15). Notice that this retains an important qualitative feature expected in the wave function. Since $\alpha_1 < \alpha_2$ for quadratic integrability, $\lambda > 1$ and A_3 is negative, as a gutter in section *a* requires. The form is now reduced to a two-parameter variation function.

† Expectation values, and the variation theorem, are discussed in *Stationary states*, Chapters 5 and 7.

Discussion 9.1

THE DISCONTINUITY AT THE RIDGE

The rotation of coordinates through 45°, from coordinates (x_1, x_2) to new coordinates (x, y) with x along the ridge (Fig. 9.6), can be embodied in a transformation:

$$x_1 = \frac{1}{\sqrt{2}}(x+y), \qquad x_2 = \frac{1}{\sqrt{2}}(-x+y). \tag{D9.1}$$

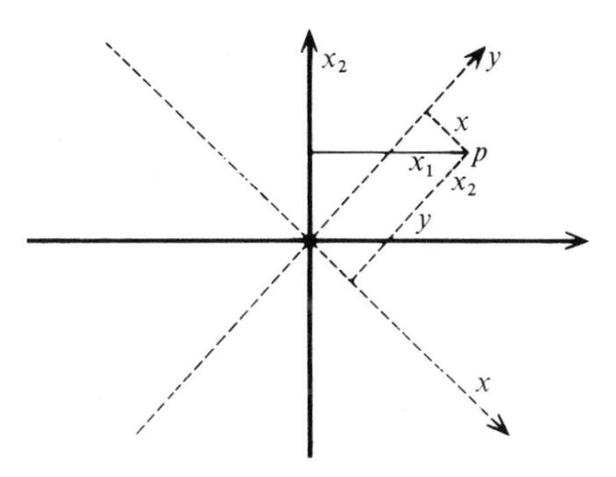

FIG. 9.6. Rotating the coordinate system from (x_1, x_2) to (x, y) is equivalent to making the transformation (D9.1).

In the new coordinate system the wave functions of interest take the forms

$$\begin{aligned}
\psi_a = & \left\{ A_1 \exp\left[(\alpha_2-\alpha_1)\frac{x}{\sqrt{2}}\right] + \right.\\
& \left. + A_2 \exp\left[(\alpha_1-\alpha_2)\frac{x}{\sqrt{2}}\right]\right\} \times \\
& \times \exp\left[-(\alpha_1+\alpha_2)\frac{y}{\sqrt{2}}\right] + \\
& + A_3 \exp\left[(\alpha_1+\alpha_2)\frac{x}{\sqrt{2}} + (\alpha_1-\alpha_2)\frac{y}{\sqrt{2}}\right], \\
\psi_c = & \left\{ A_2 \exp\left[(\alpha_2-\alpha_1)\frac{x}{\sqrt{2}}\right] + \right.\\
& \left. + A_1 \exp\left[(\alpha_1+\alpha_2)\frac{x}{\sqrt{2}}\right]\right\} \times \\
& \times \exp\left[-(\alpha_1+\alpha_2)\frac{y}{\sqrt{2}}\right] + \\
& + A_3 \exp\left[-(\alpha_1+\alpha_2)\frac{x}{\sqrt{2}} + (\alpha_1-\alpha_2)\frac{y}{\sqrt{2}}\right].
\end{aligned} \tag{D9.2}$$

The requirement on the derivatives at the ridge is

$$\left.\frac{\partial\psi_c}{\partial x}\right|_{x=0} - \left.\frac{\partial\psi_a}{\partial x}\right|_{x=0} = \eta_e \psi|_{x=0}, \tag{D9.3}$$

where η_e (read 'eta electronic') is positive and specifies the magnitude of the ridge. This equation, analogous to (9.13), places two more requirements on the A's and α's, analogous to (9.14), of which both cannot be met simultaneously by the one remaining free constant α_1.

The calculation of $\overline{H}$ is straightforward, but a little laborious. In order to make $\partial\overline{H}/\partial\alpha_1$ and $\partial\overline{H}/\partial\alpha_2$ vanish simultaneously, you must find the joint roots of two cubic expressions in α_1 and α_2. These roots furnish the minimum $\overline{H}$ that the variation function will provide.

The way to interpret the results of the procedure can be seen more simply by dropping down to a still less general variation function in which α_2 is fixed by (9.15) and only α_1 is left free. The most interesting application of the model will be to compare it with the helium atom, where a nucleus of charge $+2e$ is attracting two electrons, each of charge $-e$. Hence it is appropriate to take the depth of the well as equal to twice the height of the ridge; in other words, $\eta_n = 2\alpha_2$ (by (9.15)) and $\eta_e = \alpha_2$. Then, denoting $\mu \equiv \alpha_1/\alpha_2$,

$$\overline{H} = \frac{-3\mu^3+(\sqrt{2}+1)\mu^2+(\sqrt{2}-3)\mu-3}{3\mu+1}\alpha_2^2, \qquad (9.17)$$

and the vanishing of $\partial\overline{H}/\partial\alpha_1$ requires

$$P(\mu) \equiv -18\mu^3+(3\sqrt{2}-6)\mu^2+(2\sqrt{2}+2)\mu+\sqrt{2}+6 = 0. \qquad (9.18)$$

Searching for a root of (9.18) that lies between 0 and 1, since the wave function is significant only when $\alpha_1 < \alpha_2$, you find $\mu \doteq 0{\cdot}829$. Substituted into (9.17), this root yields

$$\overline{H} \doteq -1{\cdot}25\alpha_2^2.$$

To compare this result with an experimental value, begin by assigning a value η_n to the well that gives the right energy when there is only one electron. You know from Chapter 7 that the energy of the ground state in the hydrogenlike atom is $-E = Z^2$ atomic units, where Z is the atomic number of the nucleus. And you know from Chapter 4 that the energy of one electron in the linear delta-well model of the atom is related to η_n by $-E = \eta^2/4$. Hence for comparison with helium ($Z = 2$) it is appropriate to take $\eta_n = 4$, $\alpha_2 = 2$, providing $-\overline{H} = 1{\cdot}25\times 4 = 5$ atomic units.

The experimental ionization energy of helium—the energy required to remove one electron—is about 1·81 atomic units. After that electron is removed, the remaining electron is in the ground state of a hydrogenlike atom with $Z = 2$, and its removal will therefore require 4 atomic units of energy. Hence the total energy of the helium atom in its ground state is $-5{\cdot}81$ atomic units, to be compared with the approximation of -5. If you carry out the calculation with the two-parameter function, you will improve the approximation to $\overline{H} = -5{\cdot}21$. Of course if the electrons were independent and not repelling each other, the energy would be -8.

No doubt the three-parameter variation function would improve the agreement still further. But it would be poor taste to apply a more refined calculation to so crude a model of the atom, unless you had a reason for especial interest in the model rather than the atom.

PROBLEMS

9.1 Draw six diagrams showing the location of the nucleus and a typical location of electrons No. 1 and No. 2 corresponding to each of the six regions of Fig. 9.3.

9.2 (a) On a coordinate diagram like Fig. 9.2, draw a contour of equal probability for the wave function (9.5) for two independent electrons.
(b) Show that the shape of the contours of equal probability for the trial function (9.9) in region *b* describes a roof that is necessarily dished downwards from the ridge poles, i.e. concave upwards, when $\alpha_1 \neq \alpha_2$.

9.3 In a qualitative discussion of the wave functions for the problem treated in this chapter, comparing the properties of the 'roof' before and after pushing in the 'gutter', a recent treatise on theoretical physics says: 'The 'roof angle' at the ridges must be the same as before, since this is determined by the potential valleys.' What is wrong with this statement?

9.4 The experimental value of the energy of the lithium ion, Li^+, is $-14{\cdot}6$ atomic units. Compare this with the value given by the one-parameter variation functions of this chapter, and give the physical reason why the approximation is better or worse than the corresponding approximation for the helium atom.

9.5 Show that, for the *true* wave function for the problem of this chapter, the contributions to $\overline{H}$ from *inside* the wells and ridges must all cancel out, leaving simply the expectation value of the kinetic energy *outside* the wells and ridges (cf. Problem 4.5).

10. Many-electron Atoms

PROGRESSING through the elements in order of increasing atomic number, one finds that the increasingly numerous electrons are forced, by the properties summarized in the exclusion principle, to occupy states with different quantum numbers. Chapter 8 has shown that many conspicuous properties of an atom are determined by what the one-electron states with the highest quantum numbers are that are actually occupied by electrons when the whole atom is in its ground state. Unfortunately the many-electron problem is too complicated to treat exactly, as the hydrogenlike atom was treated in Chapter 7. It is worthwhile, nevertheless, to look further at the general problem a little more quantitatively than Chapter 8 has done, and a little less exactly than Chapter 9 has attempted.

As the atomic number of the atom increases, three major influences on the electrons change progressively. The first change is the increase in the attractive force of the nucleus because of its increasing charge. If this influence operated alone, the atoms of larger atomic number would be smaller in size than those of smaller atomic number (see Problem 7.5) because the nucleus could pull the electrons closer to it without hindrance. The second change is that of the character of the wave functions of increasing quantum number. Figs. 7.4 and 7.6 have shown that in the hydrogenlike atom the most probable location of the electron moves outward from a nucleus of any given charge as the quantum numbers increase. And finally the repulsive force that the electrons exert on one another tends to modify their wave functions so as to reduce the total electronic probability at any single place, as the simple model examined in the last chapter showed in detail.

A fairly successful way to think of the interplay of these influences is to visualize each electron as moving in a force field that is still spherically symmetrical but now composed not only of the field due to the nucleus but also of the average field due to all the remaining electrons. For an electron in such a field, the developments carried out in Chapters 5 and 6 remain valid—the angular dependences and *space quantizations* of the one-electron wave functions. Only the developments of Chapter 7, the radial dependences of the wave functions, must be modified. Now the electron moves in a field due to a point charge of $+Ze$ at the nucleus surrounded by a negative *charge cloud*, containing a total charge of $-(Z-1)e$, whose density may vary with distance from the nucleus.

To obtain a crude picture of what to expect, suppose for the moment that the charge cloud forms a ball of *uniform* density around the nucleus, a ball with a definite radius R such as Fig. 10.1 shows. Recall that the electrostatic

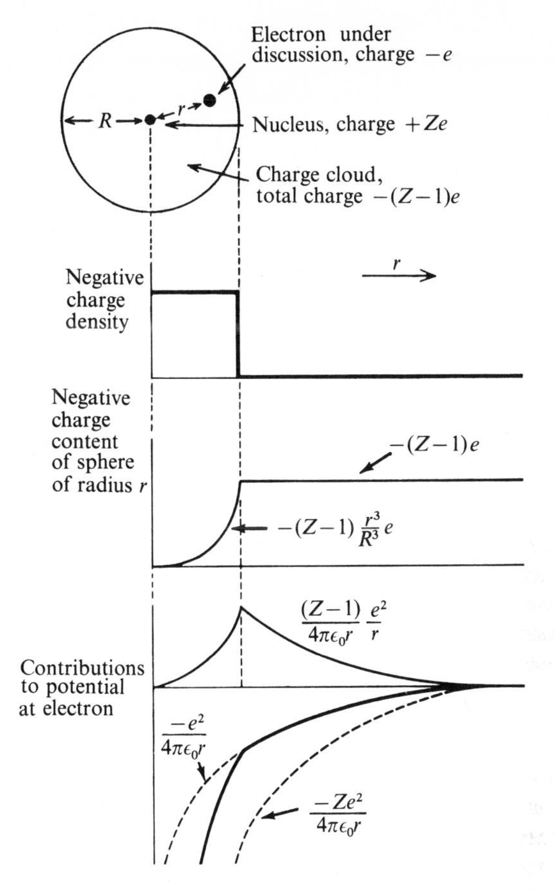

FIG. 10.1. You can think roughly of an electron as moving in the field due to the nucleus and to a spherical cloud of negative charge formed by the other electrons. The lowest solid line is the actual potential.

potential due to a thin, uniformly charged spherical shell is zero inside the shell, and that outside the shell it is the same as the potential that would arise if the charge were concentrated at the centre of the sphere. Thinking of the ball of charge as made up of concentric spherical shells, you see that the charge of the ball that is effective at the position r of the electron will be proportional to the volume of the sphere of radius r when the electron is inside the ball, and to the volume of the sphere of radius R when it is outside. Those facts prescribe the remaining plots in Fig. 10.1, showing that the potential $V(r)$ seen by the electron will vary from $-Z/r$ in atomic units when the electron is close to the nucleus to $-1/r$ when it is far from the nucleus. Hence you can think of the electron as increasingly *shielded* from the nucleus by the other electrons, as it departs increasingly from the nucleus. Furthermore, the two extremes of potential, $-Z/r$ near the nucleus and $-1/r$ far from the nucleus, will be the same whatever the variation in charge density with r, so long as the charge density falls off toward zero at large r (as it must in a bound state of the atom) and is approximately spherically symmetrical.

Notice now what will happen to the various electrons in this model of a many-electron atom. The 1s state in the hydrogenlike atom has a peak in the radial probability at $r = 1/Z$ atomic units of distance. The peaks in the

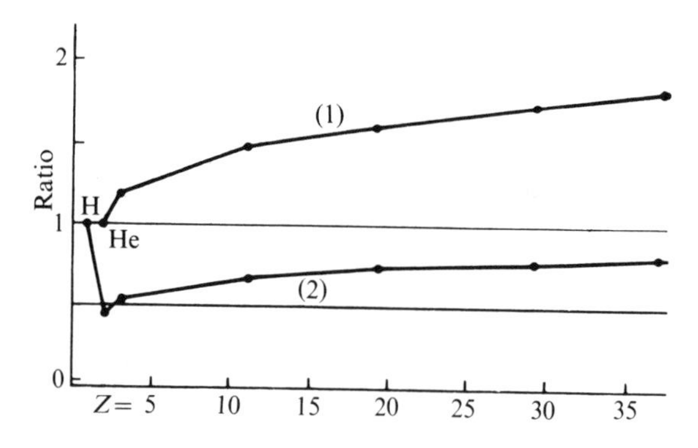

FIG. 10.2. Ratio of the energy of a 1s electron to that of (1) a 'heliumlike' electron, and (2) a 'hydrogenlike' electron.

2s and 2p hydrogenlike states are at appreciably greater distance (Figs. 7.4 and 7.6). Hence the 1s electrons will spend most of their time closer to the nucleus than the electrons of higher principal quantum number, and most of the time each will see the potential due to the nucleus modified only by that due to its companion in the 1s state of opposite spin. Their wave functions will look much like those which the two electrons in the helium atom would have if the helium nucleus had charge $+Ze$ instead of $+2e$.

Recalling that in the hydrogenlike atom the energy of an electron in the 1s state is $-Z^2$ in atomic units, you might expect that you could calculate the energy of a 1s electron when two are present in a many-electron atom, taking their interaction roughly into account, by multiplying the 1s energy for helium ($-1{\cdot}81$ atomic units) by $Z^2/2^2$. Fig. 10.2 shows the ratio of the true energy to that calculated in this way, and also the ratio to that of the hydrogenlike 1s function for the same Z, for several elements. With increasing Z the 1s electrons have energies more 'hydrogenlike' and less 'heliumlike'. You can understand this by remembering that, while the charge on the nucleus is progressively increasing, the charge on the electron is not. Thus the electronic interaction becomes a less important *fraction* of the potential for the 1s electron.

Look now at the electrons with the highest quantum numbers, for example, at the single s electron of highest principal quantum number n in the alkali metals. Even though the maximum of probability of the corresponding hydrogenlike s function is outside most of the sphere of charge of the other electrons, the s function gives a nonvanishing probability of finding the electron 'at' the nucleus. Insofar as the electron does penetrate the sphere,

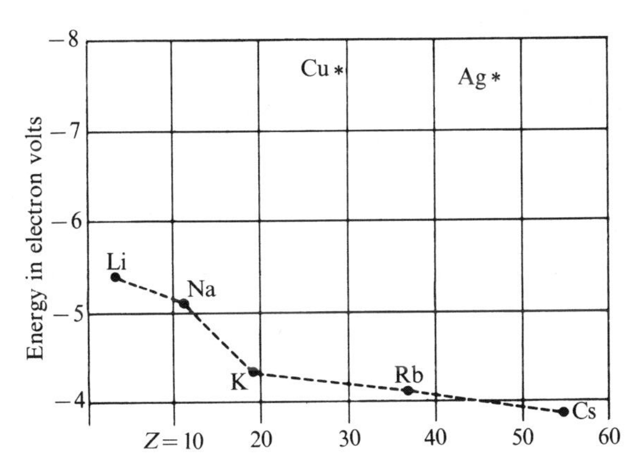

FIG. 10.3. The energies of the outermost s electron in the alkali metal atoms. Notice that on this plot an increasing ordinate represents an increasingly *negative* energy.

it will be pulled in further, since it will experience less shielding. The wave function will have some amplitude added in the region of small r, at the expense of the region of larger r; in other words, it will be smeared inward.

The energy of the electron will consequently be reduced from the value $-1/n^2$, which it would have if its wave function were hydrogenlike and the other $Z-1$ electrons shielded it so completely that it saw only the net charge $+e$. The pulling in will tend to be greater the higher the nuclear charge, reducing the energy of the electron more from the hydrogenlike value at high values of Z. There arises in this way a competition between the increased value of Z and the decreased value of $1/n^2$ as you go down the group of alkali metals. The latter influence wins, but only barely, as Fig. 10.3 shows. This figure is a plot of the minima in the ionization potential of the elements as Z changes; the argument explains the general drift downward with increasing Z that is noteworthy in Fig. 10.3. Notice also that the 4s electron in copper is at a lower energy than in the alkali metals because the 3d electrons are at a comparable distance from the nucleus much of the time and do not shield the 4s electron very effectively.

Electrons whose quantum numbers are intermediate between the lowest and the highest in the atom have intermediate energies; similar arguments can be applied to them, albeit with increased difficulty. But one other simple effect is worth notice: the effect of increasing the azimuthal quantum number l while leaving fixed the principal quantum number n. The hydrogenlike wave

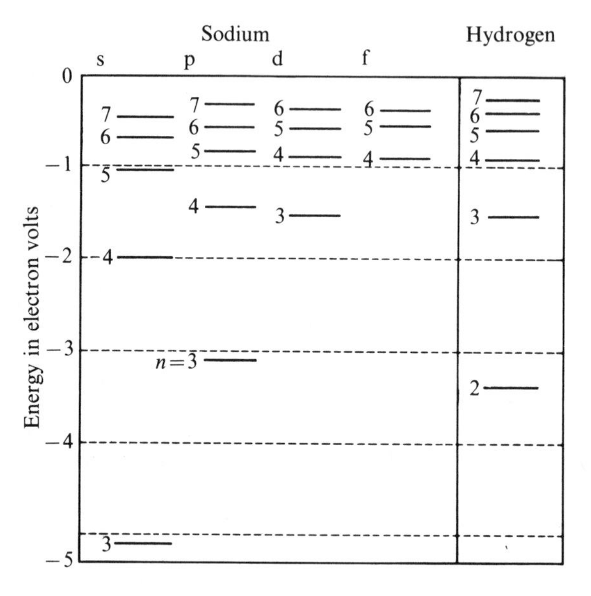

FIG. 10.4. When the 3s electron in sodium is excited to a state with a larger value of l, its energy approaches that of the corresponding level in hydrogen. Principal quantum numbers n are shown as numerals against the levels. Notice that p electrons are less tightly bound than s electrons.

functions have the factor r^l (7.8) and thus vanish at $r = 0$, unlike the s functions. Moreover that factor keeps those functions small for longer distances from the nucleus as l increases. Thus the nuclear charge has less opportunity to pull the electron toward it, and its wave function in a many-electron atom is less smeared inward, retaining more nearly a hydrogenlike form. Hence for a given value of n the energy of an electron increases with l, unlike that of the electron in the hydrogenlike atom.

The energies of the excited states of an atom demonstrate this effect in an especially clear-cut way. Fig. 10.4 compares some energy levels of the sodium atom with those of the hydrogen atom. It shows the energy levels found experimentally in a sodium atom when its 'outer' electron, normally in the 3s state whose energy level appears at the lower left, is excited to each of the other states named. As l increases, the energy of the electron rapidly approaches the level in hydrogen with the same principal quantum number. Clearly, when the remaining electron cloud is increasingly able to shield the electron from the nucleus, the electron sees more nearly the net attractive charge $+e$, and the electronic state becomes nearly the same as the excited state of hydrogen with the same quantum numbers.

Attempts to embody the foregoing considerations in a scheme of quantitative calculation take several forms. One way to obtain an approximate wave function for a one-electron state in a many-electron atom is to describe the state as hydrogenlike but with an 'effective Z', a value Z_{eff} chosen to give the experimentally observed energy in the hydrogenlike formula

$$E = -\frac{Z_{\text{eff}}^2}{n^2}. \tag{10.1}$$

Clearly Z_{eff} will always be less than the Z of the atom in question because the nucleus will always be shielded by the electron cloud. This way of thinking suggests writing

$$Z_{\text{eff}} = Z - S, \tag{10.2}$$

where S is a *shielding constant*. It then becomes tempting to regard S as a sum of shielding constants ascribed to the individual shielding electrons that make up the cloud. After evaluating S from many observations on different states of different atoms, it is possible to make up tables of individual contributions to S that give roughly consistent experimental results. It is still necessary, however, to choose such an individual shielding constant as a function not only of the quantum numbers of the shielding electron but also of the quantum numbers of the shielded electron.

A much more accurate, but a much more difficult, way of handling the problem is by the method of the *self-consistent field*. This method makes use of the fact that the electrons whose wave functions are in question, and the electrons that furnish the charge cloud, are the same electrons. If you calculate the wave function for each electron in some reasonable form of charge cloud, you can find the charge cloud that all those wave functions taken together produce. You can then repeat the calculation, using a new charge cloud, and so on until successive calculations do not change the form of the charge cloud. From an astutely chosen starting point, the process converges rapidly. It is nevertheless very laborious; for the work must all be done numerically, and relatively few atoms have been analysed by it.

Fig. 10.5(a) and (b) shows for rubidium the radial charge density thus determined—the probability of finding an electron at the distance r—for the wave functions of the inner electrons. Notice that these electrons are indeed 'inner': the high nuclear charge of $+37e$ has pulled in these functions so that the 28 electrons of $n = 1, 2$, and 3 all nearly fit into the volume of a hydrogen atom. By contrast, the 5s electron has a shallow maximum radial density at nearly six times that radius. Fig. 10.6 shows the total charge probability at the distance r.

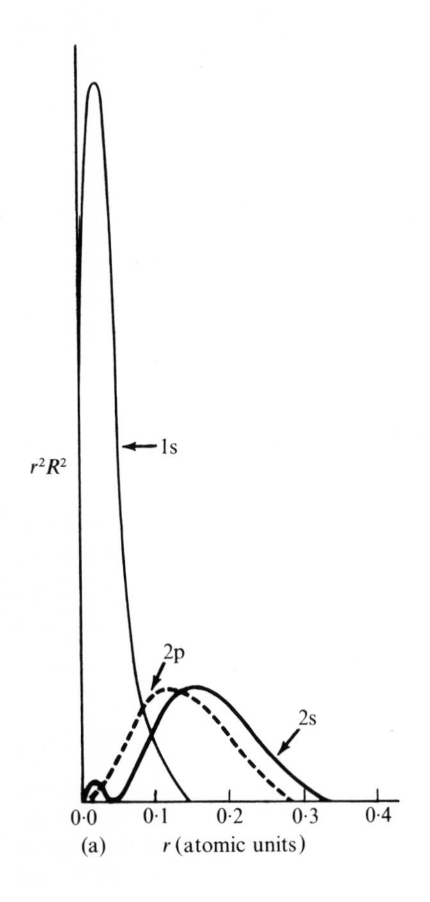

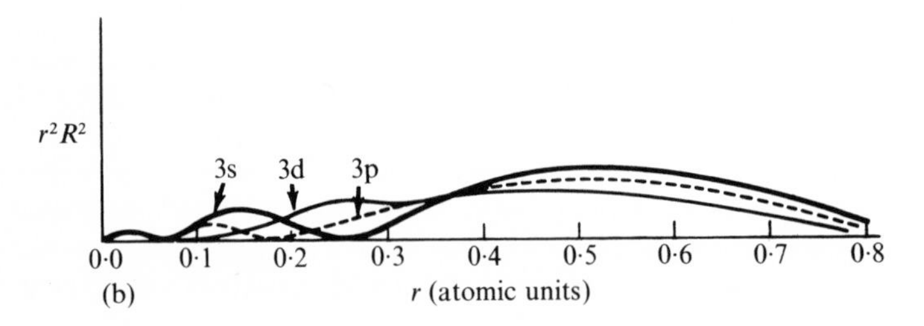

FIG. 10.5. Probability of finding an 'inner' electron at the distance r from the nucleus, in the wave functions of the states of principal quantum numbers 1, 2, and 3 in rubidium.

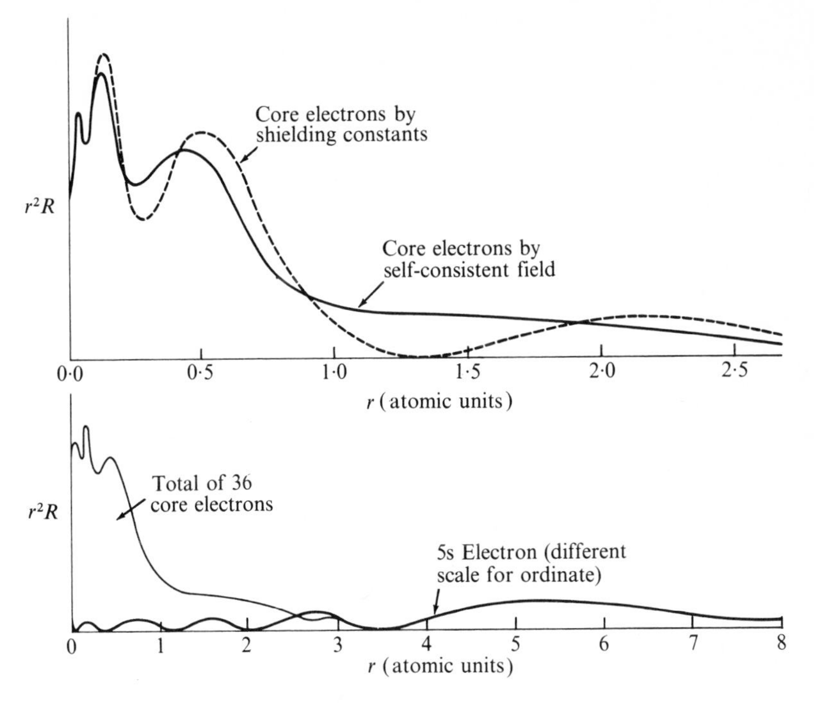

FIG. 10.6. Total charge probability at the distance r from the nucleus in rubidium.

These results confirm the pictures suggested in this chapter. The inner electrons are so tightly bound that they are only very slightly affected by the electronic behaviour at the periphery of the atom. Bonding of the atom to others, to form molecules and solids, is almost entirely controlled by the behaviour of the outer electrons. It is largely the behaviour of these electrons that the periodic table reflects. In order to disturb the inner electrons, the atom must be bombarded with quite energetic particles or with photons of short wavelength.

Examine finally the wave functions for the lowest excited states of sodium, whose energies have already been discussed and plotted in Fig. 10.4. In the hydrogenlike atom, the peaks of radial probability (the 'radii' of 'orbits') are not greatly different in all wave functions of the same principal quantum number (Figs. 7.4 and 7.6). That this is also true of wave functions for the 'inner' electrons of a many-electron atom appeared in Fig. 10.5. But Fig. 10.7(a) shows that this is not true when an outer electron is excited to a state with the same principal quantum number but with different l. The characteristic difference in penetration of the wave functions of different l, whose effect on

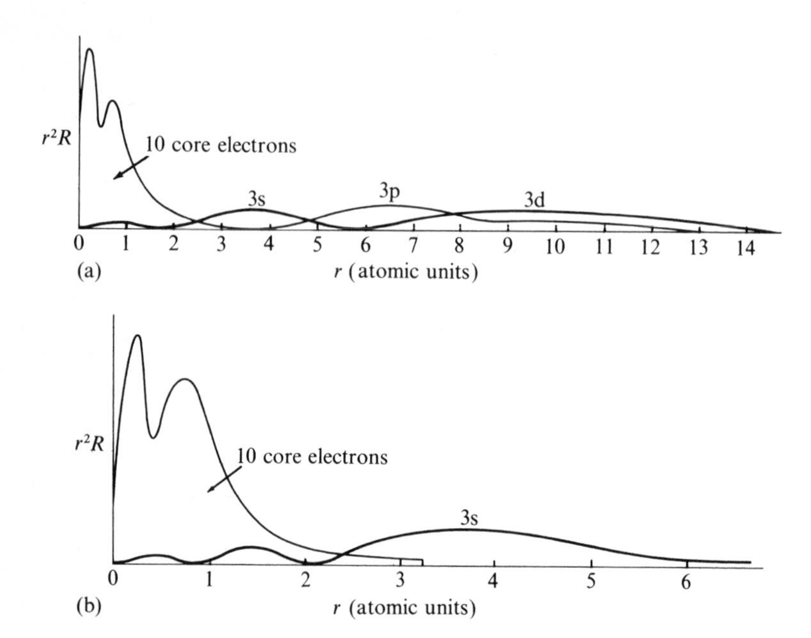

FIG. 10.7. Radial distribution of electrons in sodium: (a) ground state and first two excited states, (b) ground state.

the energies of those states appeared in Fig. 10.4, makes itself felt also on the radii. They increase with increasing l until l reaches such a value that the orbit scarcely penetrates at all; at still higher values of l the radii, like the energies, are nearly hydrogenlike.

PROBLEMS

10.1 Early in the history of atomic spectroscopy, Rydberg observed that the spectral lines from excited many-electron atoms could be correlated in a fashion we would now say was equivalent to ascribing the lines to transitions between energy levels (of excited outer electrons) given by the *Rydberg formula*

$$E = -\frac{1}{(n-\delta)^2} \text{ in atomic units,}$$

where n is the principal quantum number and δ, the so-called *quantum defect*, depends on l, the azimuthal quantum number, and is independent of n.
(a) Do you expect δ to be an increasing or a decreasing function of l?
(b) From the experimental value $E = -0{\cdot}31$ atomic units for the 5s electron of rubidium, find the value of δ for $l = 0$ in that element, use it to calculate the energy of a 4s electron (which is not an outer electron), and compare the result with the value $-2{\cdot}3$ estimated from x-ray spectroscopy.

10.2 The 'radius' of an 'orbit' in an atom, in other words the value of r at which you are most likely to find the electron, is roughly approximated by $r = n^2/(Z-S)$ atomic units, where S is a shielding constant. Find the shielding constants appropriate to the 1s, 2s, and 3s electrons in sodium whose 'orbits' have the 'radii' 0·05, 0·32, and 1·55 Ångström units.

10.3 Discuss your answer to the question: is a many-electron atom smaller or larger than the hydrogen atom would be if its electron were excited to the quantum state occupied by the outermost of the electrons in the ground state of the many-electron atom?

10.4 Offer what criticisms you can of the approximation in which you regard the electron cloud that acts upon any selected electron in an atom as spherically symmetrical (i.e. a function only of r) in the light of (1) Unsöld's Theorem (Problem 5.1), (2) the fact that only the s wave functions are spherically symmetrical, and (3) the fact that the electron repels its companions, thus digging a hole in the cloud.

TABLE 2.1 *International atomic weights 1949*

	Symbol	Atomic Number	Atomic Weight		Symbol	Atomic Number	Atomic Weight
Actinium	Ac	89	227	Neodymium	Nd	60	144·27
Aluminium	Al	13	26·97	Neon	Ne	10	20·183
Americium	Am	95	(241)	Neptunium	Np	93	(237)
Antimony	Sb	51	121·76	Nickel	Ni	28	58·69
Argon	A	18	39·944	Niobium	Nb	41	92·91
Arsenic	As	33	74·91	Nitrogen	N	7	14·008
Astatine	At	85	(210)	Osmium	Os	76	190·2
Barium	Ba	56	137·36	Oxygen	O	8	16·0000
Beryllium	Be	4	9·013	Palladium	Pd	46	106·7
Bismuth	Bi	83	209·00	Phosphorus	P	15	30·98
Boron	B	5	10·82	Platinum	Pt	78	195·23
Bromine	Br	35	79·916	Plutonium	Pu	94	(239)
Cadmium	Cd	48	112·41	Polonium	Po	84	210
Caesium	Cs	55	132·91	Potassium	K	19	39·096
Calcium	Ca	20	40·08	Praseodymium	Pr	59	140·92
Carbon	C	6	12·010	Promethium	Pm	61	(147)
Cerium	Ce	58	140·13	Protactinium	Pa	91	231
Chlorine	Cl	17	35·457	Radium	Ra	88	226·05
Chromium	Cr	24	52·01	Radon	Rn	86	222
Cobalt	Co	27	58·94	Rhenium	Re	75	186·31
Copper	Cu	29	63·54	Rhodium	Rh	45	102·91
Curium	Cm	96	(242)	Rubidium	Rb	37	85·48
Dysprosium	Dy	66	162·46	Ruthenium	Ru	44	101·7
Erbium	Er	68	167·2	Samarium	Sm	62	150·43
Europium	Eu	63	152·0	Scandium	Sc	21	45·10
Fluorine	F	9	19·00	Selenium	Se	34	78·96
Francium	Fr	87	(223)	Silicon	Si	14	28·06
Gadolinium	Gd	64	156·9	Silver	Ag	47	107·880
Gallium	Ga	31	69·72	Sodium	Na	11	22·997
Germanium	Ge	32	72·60	Strontium	Sr	38	87·63
Gold	Au	79	197·2	Sulphur	S	16	32·066
Hafnium	Hf	72	178·6	Tantalum	Ta	73	180·88
Helium	He	2	4·003	Technetium	Tc	43	(99)
Holmium	Ho	67	164·94	Tellurium	Te	52	127·61
Hydrogen	H	1	1·0080	Terbium	Tb	65	159·2
Indium	In	49	114·76	Thallium	Tl	81	204·39
Iodine	I	53	126·92	Thorium	Th	90	232·12
Iridium	Ir	77	193·1	Thulium	Tm	69	169·4
Iron	Fe	26	55·85	Tin	Sn	50	118·70
Krypton	Kr	36	83·7	Titanium	Ti	22	47·90
Lanthanum	La	57	138·92	Tungsten	W	74	183·92
Lead	Pb	82	207·21	Uranium	U	92	238·07
Lithium	Li	3	6·940	Vanadium	V	23	50·95
Lutecium	Lu	71	174·99	Xenon	Xe	54	131·3
Magnesium	Mg	12	24·32	Ytterbium	Yb	70	173·04
Manganese	Mn	25	54·93	Yttrium	Y	39	88·92
Mercury	Hg	80	200·61	Zinc	Zn	30	65·38
Molybdenum	Mo	42	95·95	Zirconium	Zr	40	91·22

TABLE 2.2 *Properties of germanium*

Mendeleev's Prediction for Es	Winkler's Observation on Ge
Atomic weight about 72	Atomic weight 72·6
Can be obtained by reduction of EsO_2 or K_2EsF_6 with Na	Obtained by reduction of K_2GeF_6 with Na
Dark grey metal of high melting temperature, relative density about 5·5	Grey metal with melting point 1231 K and relative density 5·36
On heating in air forms the oxide EsO_2 with high melting point and relative density about 4·7	Reaction with oxygen forms GeO_2 with melting point 1373 K and relative density 4·7
Forms the sulphide EsS_2, insoluble in water, soluble in ammonium sulphide solution.	GeS_2 is insoluble in water, soluble in aqueous ammonium sulphide

TABLE 2.3 *Periodic table omitting transition series*

Group		I	II	III	IV	V	VI	VII	VIII
Valency		+1	+2	+3	+4 −4	(+5) −3	(+6) −2	(+7) −1	Zero
Period	1	(1 H)						1 H	2 He
	2	3 Li	4 Be	5 B	6 C	7 N	8 O	9 F	10 Ne
	3	11 Na	12 Mg	13 Al	14 Si	15 P	16 S	17 Cl	18 A
	4	19 K	20 Ca	31 Ga	32 Ge	33 As	34 Se	35 Br	36 Kr
	5	37 Rb	38 Sr	49 In	50 Sn	51 Sb	52 Te	53 I	54 Xe
	6	55 Cs	56 Ba	81 Tl	82 Pb	83 Bi	84 Po	85 At	86 Rn

When the elements are arranged in order of increasing atomic number, the repetition of properties in successions of eight elements is made especially conspicuous by omitting certain *transition* successions of elements. The first period, containing only two elements, is rudimentary: helium might be regarded as a prototype of nonvalent elements, and hydrogen of valent elements. Periods 2 and 3 require no interruption by transition series, and exhibit almost all the important features repeated by the heavier elements. Typically a period begins with an alkali metal (strongly electropositive), reaches a halogen (strongly electronegative), and ends with a chemically inert rare gas. The elements in groups I and II form strongly basic hydroxides, those in groups V to VIII are neutral or acidic, and many of those in groups III and IV form *amphoteric* hydroxides that can act either as bases toward the later groups or as acids toward the earlier.

TABLE 2.4 *The principal transition series of elements*

Group		I	II	III	IV	V	VI	VII	VIII
Period	4	29 Cu	30 Zn	21 Sc	22 Ti	23 V	24 Cr	25 Mn	26 Fe 27 Co 28 Ni
	5	47 Ag	48 Cd	39 Y	40 Zr	41 Nb	42 Mo	43 Tc	44 Ru 45 Rh 46 Pd
	6	79 Au	80 Hg	57 La	72 Hf	73 Ta	74 W	75 Re	76 Os 77 Ir 78 Pt

The principal transition sequences, each of eight elements, interrupt periods 4, 5, and 6 of the periodic table between groups II and III. All their elements are metals, which are more or less electropositive and hence form compounds with the electronegative elements. Their group membership reflects their principal valencies (except for those placed in group VIII), but they are often *ambivalent*, exhibiting other valencies. The sixth period is itself interrupted between groups III and IV by the sequence of fourteen *rare earth* elements, sometimes called the *lanthanide* series. All the rare earths show the principal valency 3, but some can also show valency 2 or 4.

TABLE 2.5 *The periodic table to radon*

Group		I	II	III	IV	V	VI	VII	VIII
Valency		+1	+2	+3	+4 −4	(+5) −3	(+6) −2	(+7) −1	Zero
Period	1	(1 H)						1 H	2 He
	2	3 Li	4 Be	5 B	6 C	7 N	8 O	9 F	10 Ne
	3	11 Na	12 Mg	13 Al	14 Si	15 P	16 S	17 Cl	18 A
	4	19 K 29 Cu	20 Ca 30 Zn	21 Sc 31 Ga	22 Ti 32 Ge	23 V 33 As	24 Cr 34 Se	25 Mn 35 Br	26 Fe 27 Co 28 Ni 36 Kr
	5	37 Rb 47 Ag	38 Sr 48 Cd	39 Y 49 In	40 Zr 50 Sn	41 Nb 51 Sb	42 Mo 52 Te	43 Tc 53 I	44 Ru 45 Rh 46 Pd 54 Xe
	6	55 Cs 79 Au	56 Ba 80 Hg	57 La 81 Tl	72 Hf 82 Pb	73 Ta 83 Bi	74 W 84 Po	75 Re 85 At	76 Os 77 Ir 78 Pt 86 Rn

Rare earths: 58 Ce, 59 Pr, 60 Nd, 61 Pm, 62 Sm, 63 Eu, 64 Gd, 65 Tb, 66 Dy, 67 Ho, 68 Er, 69 Tm, 70 Yb, 71 Lu.

The full periodic table is here shown as far as the rare gas radon. The rare-earth sequence of 14 elements, to be inserted between atomic numbers 57 (lanthanum) and 72 (hafnium) are shown below the table. The omitted seventh period carries a few naturally occurring elements such as radium, thorium, and uranium, but all are radioactive, having more or less unstable nuclei. The elements beyond atomic number 92 (uranium) have been artifically produced only. Apparently the accumulation of more than about two hundred protons and neutrons in an atomic nucleus rapidly becomes unstable.

Index

Acidic hydroxides, 89
Alchemy, 2
Alkali metals, 8, 65, 81, 89
Alkaline earth metals, 8
Alpha particles, 5
Ambivalence, 67, 90
Amphoteric hydroxides, 89
Angular momentum, *see* Momentum
Arabian alchemy, 2
Aristotle, 2
Atom, meaning of, 3ff.
Atomic number, 11, 61, 78, 88
Atomic units, 27, 36, 52, 59, 76, 79
Atomic weight, 9, 11, 88
Atomists, Greek, 1ff.
Aufbauprinzip, 60ff.
Axial symmetry, *see* Symmetry
Azimuthal quantum number, 48, 52, 82

Basic hydroxides, 89
Bernoulli, Daniel, 3
Black-body radiation, 14
Bohr, Niels, 6, 11, 14ff., 17
Boyle, Robert, 3, 7
Bucherer, Hans, 5

Covalent bond, 65
Combining weight, 9
Carbon, 67
Cathode rays, 12
Central field, *see* Particle
Chadwick, James, 11
Chromium, 67
Coordinates
  for electron in an atom, 36, 41
  rotation of, 73, 75
Copper, 67, 81
Coulomb, Charles-Augustin, law of, 12, 17, 70

Dalton, John, 3ff., 7
Davisson, Clinton, 16
Davy, Sir Humphry, 4
De Broglie, Louis, 16, 18
Degeneracy, 40, 52ff.
Delta ridge, 70
Delta well, 29ff., 68ff.
Democritus, 1, 6
Diffuse series, 52
Direction cosines, 36
Döbereiner, Johann, 8

Eigenfunction and eigenvalue, 45, 74
Einstein, Albert, 11n., 14, 16
Eka-silicon, 10, 89
Elements,
  Aristotelian, 2
  tria prima, 3
  the modern chemical, 3, 7ff., 88
  see also Periodic table
Euler's theorem for homogeneous functions, 37, 38, 50
Excited states, 52
Exclusion principle, 23, 61, 78
Expectation value, 74ff.

Fluorine molecule, 65
Foster, Carey, 10n.
Fundamental series, 52

Gases
  kinetic theory of, 3
  rare, 7, 12
  number of molecules in, 9
Germanium, 89
Germer, Lester, 16
Greek atomistics, 1ff.
Ground state
  of particle on a line, 18
  of electron in hydrogenlike atom, 23, 52, 58
  and *aufbauprinzip*, 60

Halogens, 8, 65, 89
Harmonic oscillator, 22, 57n.
Helium
  atom, 60, 68, 76
  ion, 51, 53
Heliumlike states, 80
Hertz, Gustav, 5
Hindu philosophy, 1
Homogeneous functions, *see* Euler
Hydrogenlike wave functions, 51ff., 60ff., 80ff.
Hydroxides, 89
Hydrogen-molecule ion, 15

Ionic bond, 65ff.
Ionization, 12, 61, 64, 65, 81
Iron, 1
Isotopes, 11

Kanada, 1
Kossel, W., 16

Laguerre polynomials, 54, 58
Langmuir, Irving, 16
Lanthanide series, *see* Rare earth elements
Laplace's equation, 39
Leucippus, 1
Lewis, Gilbert, 16
Lithium, 61, 77
Living matter, atomic constituents of, 7
Lucretius, 1

Magnetic quantum number, 48
Mariotte, Edme, 3
Mendeleev, Dmitri, 10, 89
Mercury, 3
Metals, 7
Metalloids, 7, 8
Methane, 67
Meyer, Lothar, 10
Minerals, atomic constituents of, 7
Molecule, diatomic, 42
Momentum
  and wavelength, 18
  angular, of spinning electron, 23
  orbital angular, of electron in atom, 35, 44ff.
Moseley, Henry, 11

Neon, 66
Newton, Sir Isaac, 3
Neutron, 11, 90
Newlands, John, 9
Normalization, 46, 59
Nuclei, unstable atomic, 7, 90

Octaves, Newlands' law of, 10
Operators, wave-mechanical, 44, 49
Orbits, radii of, 17, 27, 59, 85, 87
Orthogonality, 41, 49, 54n.

Paracelsus, 3
Particle
  on a line, 18ff.
  in a potential well, 22ff., 25ff.
  in a central field of force, 35ff., 44ff.
  in a Coulomb field, 51ff.
Periodic table of elements
  enunciation of, 10
  shell models for, 13, 24
  and *aufbauprinzip*, 60ff.
Photoelectric effect, 14
Photon, 15, 16, 85
Planck, Max, 6, 14
Planetary model of atom, 5, 12, 17, 23
Plato, 2
Potential wells, 19, 22ff., 25ff., 35, 51, 53, 68ff., 78ff.
Principal quantum number, 52, 80ff.
Principal series, 52
Proton, 11, 90
Prout, William, 11

Quadratic integrability, 57, 73, 74
Quantum defect, 86
Quantum numbers, 48, 52, 78, 82
Quantum theory, 6, 14ff., 23, 35

Radiation by atoms, 14ff.
Radium, 5
Ramsay, Sir William, 10n.
Rare earth elements, 67, 90
Rare gases, 7, 65, 89
Rayleigh, John William Strutt, 3rd Baron, 10n.
Recursion relation, 53, 58
Rocks, atomic constituents of, 7
Rubidium, electron density in, 83
Rutherford, Sir Ernest, 5, 6, 11, 12
Rydberg constant, 17
Rydberg formula, 86
Rydberg unit of energy, 27

Salt, 1
Scandium, 62, 63
Schrödinger, Erwin, 6, 16
Schrödinger's equation
  in one dimension, 21, 26, 27
  in three dimensions, 36
  for two particles, 68ff.
Selection rules, 52
Self-consistent field calculation, 83ff.
Separation of variables, 36, 38, 42, 69
Sharp series, 52
Shells, electronic, 13, 15, 60ff.
Shielding, 80ff.
Silver, 63, 67, 81
  difluoride, 63
Sodium, excitation of, 82
Spectroscopy, 14, 16ff., 52, 86
  mass, 11
Spherical harmonics, 39ff.
Spherical symmetry, *see* Symmetry
Spinning electron, 23
Sulphur, 3
Symmetry
  of wave functions, 28ff., 71ff.
  of atom, 35, 42, 53, 78, 87
  of diatomic molecule, 42, 49

Thomson, Sir Joseph John, 5, 6, 10, 12, 15
Transition elements, 67, 89, 90
Triads of elements, Döbereiner's, 8
Tria prima, 3

Uncertainty principle, 35, 44, 45, 46
Units, atomic, 27
Unsöld's theorem, 42, 87

Valency, 9, 67, 90
Variational approximation, 74ff.
Vector-analytic notation, 42
Vibrating string, 19

Water, 1, 9
Wave mechanics, 6, 16, 21
Well, *see* Potential well
Winkler, Clemens, 10, 89